AS Maths

Edexcel Core 1

AS level maths is seriously tricky — no question about that.

We've done everything we can to make things easier for you.
We've scrutinised past paper questions and we've gone through the syllabuses with a fine-toothed comb. So we've found out exactly what you need to know, then explained it simply and clearly.

We've stuck in as many helpful hints as you could possibly want — then we even tried to put some funny bits in to keep you awake.

We've done our bit — the rest is up to you.

Contents

Section 1 — Algebra Fundamentals

A Few Definitions and Things 1
Laws of Indices 2
Surds 3
Multiplying Out Brackets 4
Taking Out Common Factors 5
Algebraic Fractions 6
Simplifying Expressions 7
Section One Revision Questions 8

Section 2 — Quadratics and Polynomials

Sketching Quadratic Graphs 9
Factorising a Quadratic 10
Completing the Square 12
The Quadratic Formula 14
Essential Proofs 16
Cows 17
Factorising Cubics 18
Section Two Revision Questions 19

Section 3 — Simultaneous Equations and Inequalities

Linear Inequalities 20
Quadratic Inequalities 21
Simultaneous Equations 22
Simultaneous Equations with Quadratics 23
Geometric Interpretation 24
Section Three Revision Questions 25

Section 4 — Coordinate Geometry and Graphs

Coordinate Geometry 26
Curve Sketching 28
Graph Transformations 29
Section Four Revision Questions 30

Section 5 — Sequences and Series

Sequences 31
Arithmetic Progressions 32
Arithmetic Series and Sigma Notation 33
Section Five Revision Questions 34

Section 6 — Differentiation

Differentiation ... 35
Finding Tangents and Normals ... 37
Section Six Revision Questions ... 38

Section 7 — Integration

Integration ... 39
Section Seven Revision Questions ... 41

P1 — Practice Exam 1

Exam ... 42
Q1 — Powers and Surds ... 44
Q2 — Inequalities and Quadratics ... 46
Q3 — Geometry ... 48
Q4 — Simultaneous Equations ... 51
Q5 — Quadratics ... 52
Q6 — Factorising and Integrating ... 54
Q7 — Equation of a Line ... 56
Q8 — Arithmetic Series ... 58
Q9 — Differentiation ... 61

P1 — Practice Exam 2

Exam ... 62
Q1 — Quadratics ... 64
Q2 — Surds ... 65
Q3 — Simultaneous Equations ... 66
Q4 — Integration and Graphs ... 68
Q5 — Algebraic Fractions ... 70
Q6 — Differentiation and Graphs ... 72
Q7 — Cubic Equations ... 75
Q8 — Indices ... 76
Q9 — Tangents and Normals ... 78
Q10 — Arithmetic Series ... 80

Answers ... 82
Index ... 84

This book covers the Core 1 module of the Edexcel specification.

Published by Coordination Group Publications Ltd.

Contributors:
Charley Darbishire
Simon Little
Andy Park
Glenn Rogers
Claire Thompson

And:
Iain Nash

Updated by:
Alison Chisholm
Tim Major
Sam Norman
Andy Park
Alan Rix
Alice Shepperson
Claire Thompson
Julie Wakeling

ISBN: 1-84146-765-0

Groovy website: www.cgpbooks.co.uk

Jolly bits of clipart from CorelDRAW
With thanks to Colin Wells and Vicky Daniel for the proofreading.

Printed by Elanders Hindson, Newcastle upon Tyne.

A Few Definitions and Things

Yep, this is a pretty dull way to start a book. A list of definitions. But in the words of that annoying one-hit wonder bloke in the tartan suit, things can only get better. Which is nice.

Polynomials

POLYNOMIALS are expressions of the form $a + bx + cx^2 + dx^3 + ...$

$5y^3 + 2y + 23$ ← Polynomial in the variable y.

$1 + x^2$

$z^{42} + 3z - z^2 - 1$ ← Polynomial in the variable z.

The bits separated by the +/– signs are terms.

x, y and z are always VARIABLES. They're usually what you solve equations to find. They often have more than one possible value.

Letters like a, b, c are always CONSTANTS. Constants never change. They're fixed numbers — but can be represented by letters. π is a good example. You use the symbol π, but it's just a number = 3.1415...

Functions

FUNCTIONS take a value, do something to it, and output another value.

$f(x) = x^2 + 1$ ← function f takes a value, squares it and adds 1.

$g(x) = 2 - \sin 2x$ ← function g takes a value (in degrees), doubles it, takes the sine of it, then takes the value away from 2.

You can plug values into a function — just replace the variable with a certain number.

$f(-2) = (-2)^2 + 1 = 5$

$f(0) = (0)^2 + 1 = 1$

$f(252) = (252)^2 + 1 = 63505$

$g(-90) = 2 - \sin(-180°) = 2 - 0 = 2$

$g(0) = 2 - \sin 0° = 2 - 0 = 2$

$g(45) = 2 - \sin 90° = 2 - 1 = 1$

Exam questions use functions all the time. They generally don't have that much to do with the actual question. It's just a bit of terminology to get comfortable with.

Multiplication and Division

There's three different ways of showing MULTIPLICATION:

1) with good old-fashioned "times" signs (×):

$f(x) = (2x \times 6y) + (2x \times \sin x) + (z \times y)$ ← The multiplication signs and the variable x are easily confused.

2) or sometimes just use a little dot:

$f(x) = 2x.6y + 2x.\sin x + z.y$ ← Dots are better for long expressions — they're less confusing and easier to read.

3) but you often don't need anything at all:

$f(x) = 12xy + 2x \sin x + zy$

And there's three different ways of showing DIVISION:

1) $\frac{x+2}{3}$

2) $(x+2) \div 3$

3) $(x+2)/3$

Equations and Identities

This is an IDENTITY:

$$x^2 - y^2 \equiv (x+y)(x-y)$$

Make up any values you like for x and y, and it's always true. The left-hand side always equals the right-hand side.

But this is an EQUATION:

$$y = x^2 + x$$

This has at most two possible solutions for each value of y. e.g. if y=0, x can only be 0 or -1.

The difference is that the identity's true for all values of x and y, but the equation's only true for certain values.

NB: If it's an identity, use the ≡ sign instead of =.

Laws of Indices

You use the laws of indices a helluva lot in maths — when you're integrating, differentiating and ...er... well loads of other places. So take the time to get them sorted now.

Three mega-important Laws of Indices

You must know these three rules. I can't make it any clearer than that.

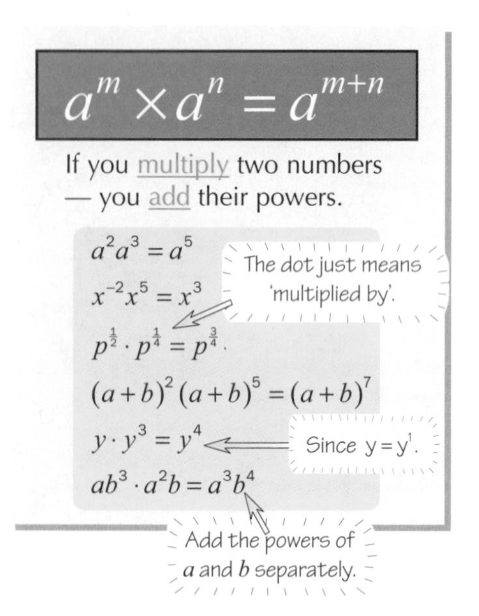

$$a^m \times a^n = a^{m+n}$$

If you multiply two numbers — you add their powers.

$a^2a^3 = a^5$

$x^{-2}x^5 = x^3$

$p^{\frac{1}{2}} \cdot p^{\frac{1}{4}} = p^{\frac{3}{4}}$

$(a+b)^2(a+b)^5 = (a+b)^7$

$y \cdot y^3 = y^4$

$ab^3 \cdot a^2b = a^3b^4$

$$\frac{a^m}{a^n} = a^{m-n}$$

If you divide two numbers — you subtract their powers.

$\frac{x^5}{x^2} = x^3$

$\frac{x^{\frac{3}{4}}}{x} = x^{-\frac{1}{4}}$

$\frac{x^3y^2}{xy^3} = x^2y^{-1}$

Subtract the powers of x and y separately.

$$\left(a^m\right)^n = a^{mn}$$

If you have a power to the power of something else — multiply the powers together.

$(x^2)^3 = x^6$

$\left\{(a+b)^3\right\}^4 = (a+b)^{12}$

$(ab^2)^4 = a^4\left(b^2\right)^4 = a^4b^8$

This power applies to both bits inside the brackets.

Other important stuff about Indices

You can't get very far without knowing this sort of stuff. Learn it — you'll definitely be able to use it.

$$a^{\frac{1}{m}} = \sqrt[m]{a}$$

You can write roots as powers...

EXAMPLES:

$x^{\frac{1}{5}} = \sqrt[5]{x}$

$4^{\frac{1}{2}} = \sqrt{4} = 2$

$125^{\frac{1}{3}} = \sqrt[3]{125} = 5$

$$a^{\frac{m}{n}} = \sqrt[n]{a^m} = \left(\sqrt[n]{a}\right)^m$$

A power that's a fraction like this is the root of a power — or the power of a root.

EXAMPLES:

$9^{\frac{3}{2}} = \left(9^{\frac{1}{2}}\right)^3 = \left(\sqrt{9}\right)^3 = 3^3 = 27$

$16^{\frac{3}{4}} = \left(16^{\frac{1}{4}}\right)^3 = \left(\sqrt[4]{16}\right)^3 = 2^3 = 8$

It's often easier to work out the root first, then raise it to the power.

$$a^{-m} = \frac{1}{a^m}$$

A negative power means it's on the bottom line of a fraction.

EXAMPLES:

$x^{-2} = \frac{1}{x^2}$

$2^{-3} = \frac{1}{2^3} = \frac{1}{8}$

$(x+1)^{-1} = \frac{1}{x+1}$

$$a^0 = 1$$

This works for any number or letter.

EXAMPLES:

$x^0 = 1$

$2^0 = 1$

$(a+b)^0 = 1$

Indices, indices — de fish all live indices...

What can I say that I haven't said already? Blah, blah, important. Blah, blah, learn these. Blah, blah, use them all the time. Mmm, that's about all that needs to be said really. So I'll be quiet and let you get on with what you need to do.

Surds

A surd is a number like $\sqrt{2}$, $\sqrt[3]{12}$ or $5\sqrt{3}$ — one that's written with the $\sqrt{\ }$ sign. They're important because you can give exact answers where you'd otherwise have to round to a certain number of decimal places.

Surds are sometimes the only way to give an Exact Answer

Put $\sqrt{2}$ into a calculator and you'll get something like 1.414213562...
But square 1.414213562 and you get 1.999999999.

And no matter how many decimal places you use, you'll never get exactly 2.
The only way to write the exact, spot on value is to use surds.

So unless the question asks for an answer to a certain number of decimal places — leave your answer as a surd.

There are basically Three Rules for using Surds

There are three rules you'll need to know to be able to use surds properly. Check out the 'Rules of Surds' box below.

EXAMPLES: (i) Simplify $\sqrt{12}$ and $\sqrt{\frac{3}{16}}$. (ii) Show that $\frac{9}{\sqrt{3}} = 3\sqrt{3}$. (iii) Find $\left(2\sqrt{5}+3\sqrt{6}\right)^2$.

(i) Simplifying surds means making the number in the $\sqrt{\ }$ sign smaller, or getting rid of a fraction in the $\sqrt{\ }$ sign.

$$\sqrt{12} = \sqrt{4\times3} = \sqrt{4}\times\sqrt{3} = 2\sqrt{3}$$

Using $\sqrt{ab} = \sqrt{a}\sqrt{b}$.

$$\sqrt{\frac{3}{16}} = \frac{\sqrt{3}}{\sqrt{16}} = \frac{\sqrt{3}}{4}$$

Using $\sqrt{\frac{a}{b}} = \frac{\sqrt{a}}{\sqrt{b}}$.

(ii) For questions like these, you have to write a number (here, it's 3) as $3 = \left(\sqrt{3}\right)^2 = \sqrt{3}\times\sqrt{3}$.

$$\frac{9}{\sqrt{3}} = \frac{3\times3}{\sqrt{3}} = \frac{3\times\sqrt{3}\times\sqrt{3}}{\sqrt{3}} = 3\sqrt{3}$$

Cancelling $\sqrt{3}$ from the top and bottom lines.

(iii) Multiply surds very carefully — it's easy to make a silly mistake.

$$\begin{aligned}\left(2\sqrt{5}+3\sqrt{6}\right)^2 &= \left(2\sqrt{5}+3\sqrt{6}\right)\left(2\sqrt{5}+3\sqrt{6}\right)\\ &= \left(2\sqrt{5}\right)^2 + 2\times\left(2\sqrt{5}\right)\times\left(3\sqrt{6}\right) + \left(3\sqrt{6}\right)^2\\ &= \left(2^2\times\sqrt{5}^2\right) + \left(2\times2\times3\times\sqrt{5}\times\sqrt{6}\right) + \left(3^2\times\sqrt{6}^2\right)\\ &= 20 + 12\sqrt{30} + 54\\ &= 74 + 12\sqrt{30}\end{aligned}$$

$= 4\times5 = 20$ $\quad$ $= 12\sqrt{5}\sqrt{6} = 12\sqrt{30}$ $\quad$ $= 9\times6 = 54$

Rules of Surds

There's not really very much to remember.

$$\sqrt{ab} = \sqrt{a}\sqrt{b}$$

$$\sqrt{\frac{a}{b}} = \frac{\sqrt{a}}{\sqrt{b}}$$

$$a = \left(\sqrt{a}\right)^2 = \sqrt{a}\sqrt{a}$$

Remove surds from fractions by Rationalising the Denominator

Surds are pretty darn complicated.

So they're the last thing you want at the bottom of a fraction.

But have no fear — Rationalise the Denominator...

Yup, you heard... (it means getting rid of the surds from the bottom of a fraction).

EXAMPLE: Rationalise the denominator of $\frac{1}{1+\sqrt{2}}$

Multiply the top and bottom by the denominator (but change the sign in front of the surd).

$$\frac{1}{1+\sqrt{2}} \times \frac{1-\sqrt{2}}{1-\sqrt{2}}$$

$$\frac{1-\sqrt{2}}{\left(1+\sqrt{2}\right)\left(1-\sqrt{2}\right)} = \frac{1-\sqrt{2}}{1^2+\sqrt{2}-\sqrt{2}-\sqrt{2}^2}$$

This works because: $(a+b)(a-b) = a^2 - b^2$

$$\frac{1-\sqrt{2}}{1-2} = \frac{1-\sqrt{2}}{-1} = -1+\sqrt{2}$$

Surely the pun is mightier than the surd...

There's not much to surds really — but they cause a load of hassle. Think of them as just ways to save you the bother of getting your calculator out and pressing buttons — then you might grow to know and love them. The box of rules in the middle is the vital stuff. Learn them till you can write them down without thinking — then get loads of practice with them.

Multiplying Out Brackets

In this horrific nightmare that is AS-level maths, you need to manipulate and simplify expressions all the time.

Remove brackets by Multiplying them out

Here are the basic types you have to deal with. You'll have seen them before. But there's no harm in reminding you, eh?

Multiply Your Brackets Here — we do all shapes and sizes

Single Brackets

$a(b+c+d)=ab+ac+ad$

Double Brackets

$(a+b)(c+d)=ac+ad+bc+bd$

Long Brackets

Write it out again with each term from one bracket separately multiplied by the other bracket.

$(x+y+z)(a+b+c+d)$

$=x(a+b+c+d)+y(a+b+c+d)+z(a+b+c+d)$

Then multiply out each of these brackets, one at a time.

Squared Brackets

$(a+b)^2=(a+b)(a+b)=a^2+2ab+b^2$

Use the middle stage until you're comfortable with it. Just never make this mistake: $(a+b)^2=a^2+b^2$

Single Brackets

$3xy(x^2+2x-8)$ — Multiply all the terms inside the brackets by the bit outside — separately.

$(3xy\times x^2)+(3xy\times 2x)+(3xy\times(-8))$ — I've put brackets round each bit to make it easier to read.

All the stuff in the brackets now needs sorting out. Work on each bracket separately.

$(3x^3y)+(6x^2y)+(-24xy)$ — Multiply the numbers first, then put the x's and other letters together.

$3x^3y+6x^2y-24xy$

Squared Brackets

$(2y^2+3x)^2$

Either write it as two brackets and multiply it out...

$(2y^2+3x)(2y^2+3x)$

$2y^2.2y^2+2y^2.3x+3x.2y^2+3x.3x$ — The dot just means 'multiplied by' — the same as the × sign.

$4y^4+6xy^2+6xy^2+9x^2$ — From here on it's simplification — nothing more, nothing less.

$4y^4+12xy^2+9x^2$

...or do it in one go.

$(2y^2)^2+2(2y^2)(3x)+(3x)^2$

a^2 $2ab$ b^2

$4y^4+12xy^2+9x^2$

Long Brackets

$(2x^2+3x+6)(4x^3+6x^2+3)$

$2x^2(4x^3+6x^2+3)+3x(4x^3+6x^2+3)+6(4x^3+6x^2+3)$ — Each term in the first bracket has been multiplied by the second bracket.

$(8x^5+12x^4+6x^2)+(12x^4+18x^3+9x)+(24x^3+36x^2+18)$ — Now multiply out each of these brackets.

$8x^5+24x^4+42x^3+42x^2+9x+18$ — Then simplify it all...

Go forth and multiply out brackets...

OK, so this is obvious, but I'll say it anyway — if you've got 3 or more brackets together, multiply them out 2 at a time. Then you'll be turning a really hard problem into two easy ones. You can do that loads in maths. In fact, writing the same thing in different ways is what maths is about. That and sitting in classrooms with tacky 'maths can be fun' posters...

Taking Out Common Factors

Common factors need to be hunted down, and taken outside the brackets. They are a danger to your exam mark.

Spot those Common Factors

A bit which is in each term of an expression is a common factor.

Spot Those Common Factors $2x^3z + 4x^2yz + 14x^2y^2z$

Look for any bits that are in each term.

Numbers: there's a common factor of 2 here because 2 divides into 2, 4 and 14.

Variables: there's at least an x^2 in each term and there's a z in each term.

So there's a common factor of $2x^2z$ in this expression.

And Take Them Outside a Bracket

If you spot a common factor you can "take it out":

$$2x^2z(x+2y+7y^2)$$

Write the common factor outside a bracket.

and put what's left of each term inside the bracket.

Afterwards, always multiply back out to check you did it right:

Check by Multiplying Out Again

$$2x^2z(x+2y+7y^2) = 2x^3z + 4x^2yz + 14x^2y^2z$$

But it's not just numbers and variables you need to look for...

Trig Functions: $\sin x \sin y + \cos x \sin y$

This has a common factor of sin y. So take it out to get...

$$\sin y(\sin x + \cos x)$$

Brackets: $(y+a)^2(x-a)^3 + (x-a)^2$

$(x - a)^2$ is a common factor — it comes out to give:

$$(x-a)^2\left((y+a)^2(x-a)+1\right)$$

Look for Common Factors when Simplifying Expressions

EXAMPLE: Simplify... $(x+1)(x-2)+(x+1)^2 - x(x+1)$

There's an (x+1) factor in each term, so we can take this out as a common factor (hurrah).

$$(x+1)\{(x-2)+(x+1)-x\}$$

The terms inside the big bracket are the old terms with an (x+1) removed.

At this point you should check that this multiplies out to give the original expression. (You can just do this in your head, if you trust it.)

Then simplify the big bracket's innards:

$$(x+1)(\cancel{x}-2+x+1-\cancel{x})$$

$$=(x+1)(x-1)$$

$$= x^2-1$$

Get this answer by multiplying out the two brackets (or by using the "difference of two squares").

Bored of spotting trains or birds? Try common factors...

You'll be doing this business of taking out common factors a lot — so get your head round this. It's just a case of looking for things that are in all the different terms of an expression, i.e. bits they have in common. And if something's in all the different terms, save yourself some time and ink, and write it once — instead of two, three or more times.

Algebraic Fractions

No one likes fractions. But just like Mondays, you can't put them off forever. Face those fears. Here goes...

The first thing you've got to know about fractions:

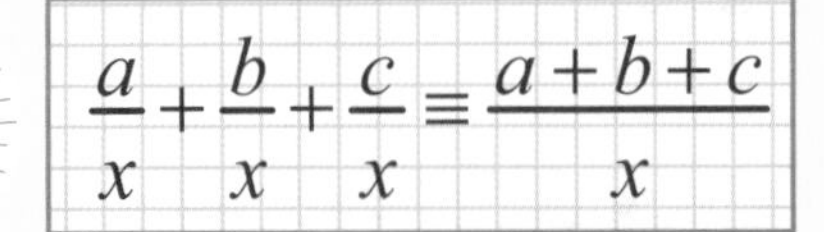

$$\frac{a}{x}+\frac{b}{x}+\frac{c}{x}\equiv\frac{a+b+c}{x}$$

You can just add the stuff on the top lines because the bottom lines are all the same.

x is called a common denominator — a fancy way of saying 'the bottom line of all the fractions is x'.

Add fractions by putting them over a **Common Denominator...**

Finding a common denominator just means 'rewriting some fractions so all their bottom lines are the same'.

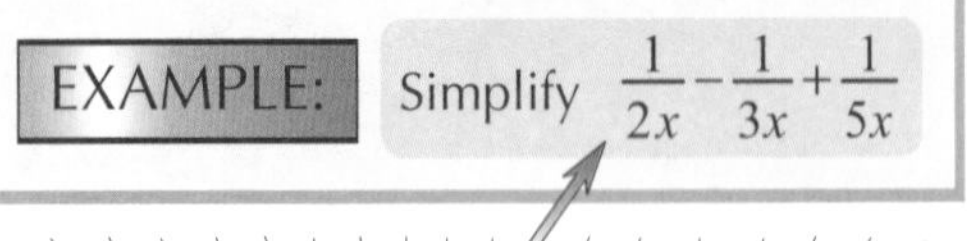

EXAMPLE: Simplify $\frac{1}{2x}-\frac{1}{3x}+\frac{1}{5x}$

You need to rewrite these so that all the bottom lines are equal. What you want is something that all these bottom lines divide into.

Put It over a Common Denominator

30 is the lowest number that 2, 3, and 5 go into.
So the common denominator is 30x.

$$\frac{15}{30x}-\frac{10}{30x}+\frac{6}{30x}$$

Always check that these divide out to give what you started with.

$$=\frac{15-10+6}{30x}=\frac{11}{30x}$$

...even **horrible** looking ones

Yep, finding a common denominator even works for those fraction nasties — like these:

EXAMPLE: Find $\frac{2y}{x(x+3)}+\frac{1}{y^2(x+3)}-\frac{x}{y}$

Find the Common Denominator

Take all the individual 'bits' from the bottom lines and multiply them together.
Only use each bit once unless something on the bottom line is squared.

The individual 'bits' here are x, (x+3) and y...

$$xy^2(x+3)$$

...but you need to use y^2 because there's a y^2 in the second fraction's denominator.

Put Each Fraction over the Common Denominator

Make the denominator of each fraction into the common denominator.

$$\frac{y^2\times 2y}{y^2x(x+3)}+\frac{x\times 1}{xy^2(x+3)}-\frac{xy(x+3)\times x}{xy(x+3)y}$$

Multiply the top and bottom lines of each fraction by whatever makes the bottom line the same as the common denominator.

Combine into One Fraction

Once everything's over the common denominator — you can just add the top lines together.

$$=\frac{2y^3+x-x^2y(x+3)}{xy^2(x+3)}$$

As always — if you see a minus sign, look out for possible problems.

All the bottom lines are the same — so you can just add the top lines.

$$=\frac{2y^3+x-x^3y-3x^2y}{xy^2(x+3)}$$

All you need to do now is tidy up the top.

Not the nicest of answers. But it <u>is</u> the answer, so it'll have to do.

Well put me over a common denominator and pickle my walrus...

Adding fractions — turning lots of fractions into one fraction. Sounds pretty good to me, since it means you don't have to write as much. Better do it carefully, though — otherwise you can watch those marks shoot straight down the toilet.

Simplifying Expressions

I know this is basic stuff but if you don't get really comfortable with it you will make silly mistakes. You will.

Cancelling stuff on the top and bottom lines

Cancelling stuff is good — because it means you've got rid of something, and you don't have to write as much.

EXAMPLE: Simplify $\frac{ax+ay}{az}$

You can do this in two ways. Use whichever you prefer — but make sure you understand the ideas behind both.

Factorise — then Cancel

$$\frac{ax+ay}{az}=\frac{a(x+y)}{az}$$

Factorise the top line.

$$=\frac{\not{a}(x+y)}{\not{a}z}=\frac{x+y}{z}$$

Cancel the 'a'.

Split into Two Fractions — then Cancel

$$\frac{ax+ay}{az}=\frac{ax}{az}+\frac{ay}{az}$$

This is an okay thing to do — just think what you'd get if you added these.

$$=\frac{\not{a}x}{\not{a}z}+\frac{\not{a}y}{\not{a}z}=\frac{x}{z}+\frac{y}{z}$$

This answer's the same as the one from the first box — honest. Check it yourself by adding the fractions.

Simplifying complicated-looking Brackets

EXAMPLE: Simplify the expression $(x-y)(x^2+xy+y^2)$

There's only one thing to do here.... Multiply out those brackets!

$$(x-y)(x^2+xy+y^2)=x(x^2+xy+y^2)-y(x^2+xy+y^2)$$

Multiplying each term in the first bracket by the second bracket.

$$=(x^3+x^2y+xy^2)-(x^2y+xy^2+y^3)$$

Multiplying out each of these two brackets.

$$=x^3+x^2y+xy^2-x^2y-xy^2-y^3$$

Don't forget these become minus signs because of the minus sign in front of the bracket.

And then the x^2y and the xy^2 terms disappear...

$$=x^3-y^3$$

Sometimes you just have to do Anything you can think of and Hope...

Sometimes it's not easy to see what you're supposed to do to simplify something.
When this happens — just do anything you can think of and see what 'comes out in the wash'.

EXAMPLE: Simplify $4x+\frac{4x}{x+1}-4(x+1)$

There's nothing obvious to do — so do what you can. Try adding them as fractions...

$$4x+\frac{4x}{x+1}-4(x+1)=\frac{(x+1)\times 4x}{x+1}+\frac{4x}{x+1}-\frac{(x+1)\times 4(x+1)}{x+1}$$

The common denominator is $(x + 1)$.

$$=\frac{4x^2+4x+4x-4(x+1)^2}{x+1}$$

Still looks horrible. So work out the brackets — but don't forget the minus signs.

$$=\frac{4x^2+4x+4x-4x^2-8x-4}{x+1}$$

$$=-\frac{4}{x+1}$$

Aha — everything disappears to leave you with this. And this is definitely simpler than it looked at the start.

Don't look at me like that...

Choose a word, any word at all. Like "Simple". Now stare at it. Keep staring at it. Does it look weird? No? Stare a bit longer. Now does it look weird? Yes? Why is that? I don't understand.

Section One Revision Questions

So that was the first section. And let's face it, it wasn't that bad. But it's all really important basic stuff that you need to be very comfortable with — otherwise AS-level maths will just become a never-ending misery. Anyway, before you get stuck into Section Two, test yourself with these questions. Go on. If you thought this section was a doddle, you should be able fly through them... (They're also very helpful if you're having trouble sleeping.)

1) Pick out the constants and the variables from the following equations:

   a) $(ax+6)^2 = 2b+3$ b) $\sin k\theta = 5$ c) $y = \frac{-b \pm \sqrt{b^2-4ac}}{2a}$ d) $\frac{dy}{dx} = x^2+ax+2$

2) What symbol should be used instead of the equals sign in identities?

3) Which of these are identities (i.e. true for all variable values)?

   A $(x+b)(y-b) = xy + b(y-x) - b^2$ B $(2y+x)^2 = 10$

   C $\tan\theta = \frac{\sin\theta}{\cos\theta}$ D $a^3+b^3 = (a+b)(a^2-ab+b^2)$

4) Number 6 in my all-time top ten functions is: f defined by $f(x)= (x+1)^2/3x$. Find the value (to 1 dp) of f when...

   a) $x = -5$ b) $x = 0$ c) $x = 25$ d) $x = -1$

5) Simplify these:

   a) $x^3.x^5$ b) $a^7.a^8$ c) $\frac{x^8}{x^2}$ d) $(a^2)^4$ e) $(xy^2).(x^3yz)$ f) $\frac{a^2b^4c^6}{a^3b^2c}$

6) Work out the following:

   a) $16^{\frac{1}{2}}$ b) $8^{\frac{1}{3}}$ c) $16^{\frac{3}{4}}$ d) x^0 e) $49^{-\frac{1}{2}}$

7) Find exact answers to these equations:

   a) $x^2 - 5 = 0$ b) $(x+2)^2 - 3 = 0$

8) Simplify:

   a) $\sqrt{28}$ b) $\sqrt{\frac{5}{36}}$ c) $\sqrt{18}$ d) $\sqrt{\frac{9}{16}}$

9) Show that a) $\frac{8}{\sqrt{2}} = 4\sqrt{2}$, and b) $\frac{\sqrt{2}}{2} = \frac{1}{\sqrt{2}}$

10) Find $(6\sqrt{3}+2\sqrt{7})^2$

11) Rationalise the denominator of: $\frac{2}{3+\sqrt{7}}$

12) Remove the brackets and simplify the following expressions:

    a) $(a+b)(a-b)$ b) $(a+b)(a+b)$

    c) $35xy + 25y(5y+7x) - 100y^2$ d) $(x+3y+2)(3x+y+7)$

13) Take out the common factors from the following expressions:

    a) $2x^2y + axy + 2xy^2\sin x$ b) $\sin^2x + \cos^2x\sin^2x$ c) $16y + 8yx + 56x$ d) $x(x-2)+3(2-x)$

14) Put the following expressions over a common denominator:

    a) $\frac{2x}{3} + \frac{y}{12} + \frac{x}{5}$ b) $\frac{5}{xy^2} - \frac{2}{x^2y}$ c) $\frac{1}{x} + \frac{x}{x+y} + \frac{y}{x-y}$

15) Simplify these expressions:

    a) $\frac{2a}{b} - \frac{a}{2b}$ b) $\frac{2p}{p+q} + \frac{2q}{p-q}$ c) "A bird in the hand is worth two in the bush"

Sketching Quadratic Graphs

If a question doesn't seem to make sense, or you can't see how to go about solving a problem, try drawing a graph. It sometimes helps if you can actually see what the problem is, rather than just reading about it.

Sketch the graphs of the following quadratic functions:

① $y = 2x^2 - 4x + 3$

② $y = 8 - 2x - x^2$

Quadratic graphs are *Always* u-shaped or n-shaped

A The first thing you need to know is whether the graph's going to be u-shaped or n-shaped (upside down). To decide, look at the coefficient of x^2.

$y = 2x^2 - 4x + 3$

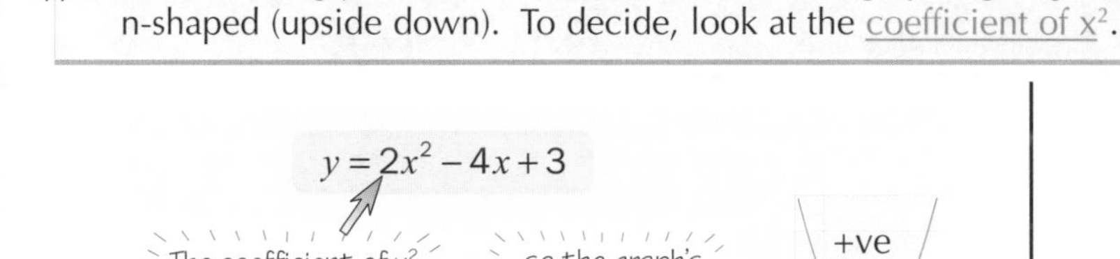

$y = 8 - 2x - x^2$

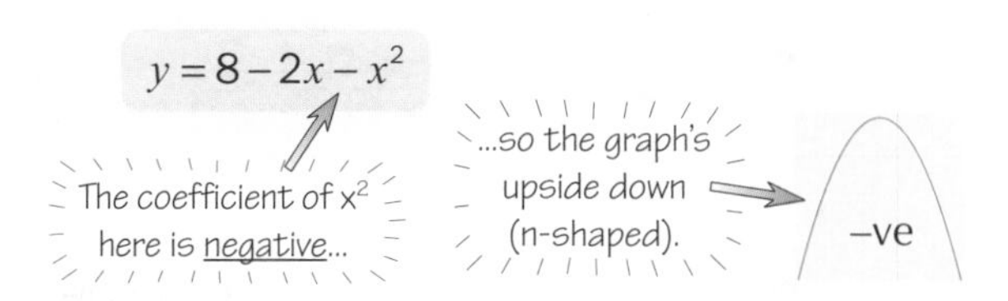

B Now find the places where the graph crosses the axes (both the y-axis and the x-axis).

(i) Put x=0 to find where it meets the y-axis.

$y = 2x^2 - 4x + 3$

$y = (2\times0^2) - (4\times0) + 3$ so $y = 3$ ← That's where it crosses the y-axis

(ii) Solve y=0 to find where it meets the x-axis.

$2x^2 - 4x + 3 = 0$

$b^2 - 4ac = -8 < 0$ ← You could use the formula. But first check $b^2 - 4ac$ to see if y = 0 has any roots.

So it has no solutions, and doesn't cross the x-axis.

For more info, see page 15.

(i) Put x=0.

$y = 8 - 2x - x^2$

$y = 8 - (2\times0) - 0^2$ so $y = 8$

(ii) Solve y=0.

$8 - 2x - x^2 = 0$ ← This equation factorises easily...

$\Rightarrow (2 - x)(x + 4) = 0$

$\Rightarrow x = 2$ or $x = -4$

C Finally, find the minimum or maximum (i.e. the vertex).

Since $y = 2(x-1)^2 + 1$ ← By completing the square (see page 12).

the minimum value is $y = 1$, which occurs at $x = 1$

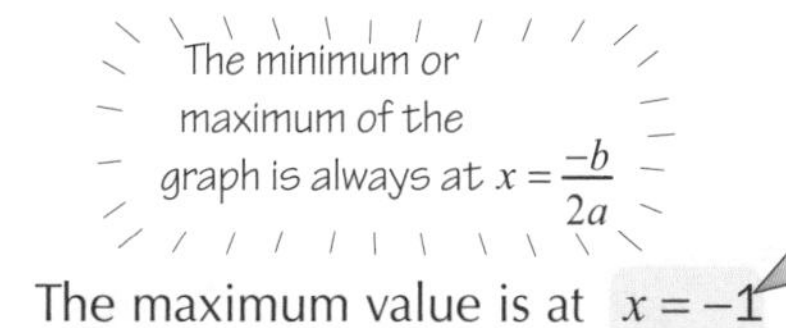

The maximum value is halfway between the roots — the graph's symmetrical.

The maximum value is at $x = -1$

So the maximum is $y = 8 - (2\times-1) - (-1)^2$

i.e. the graph has a maximum at the point (–1,9).

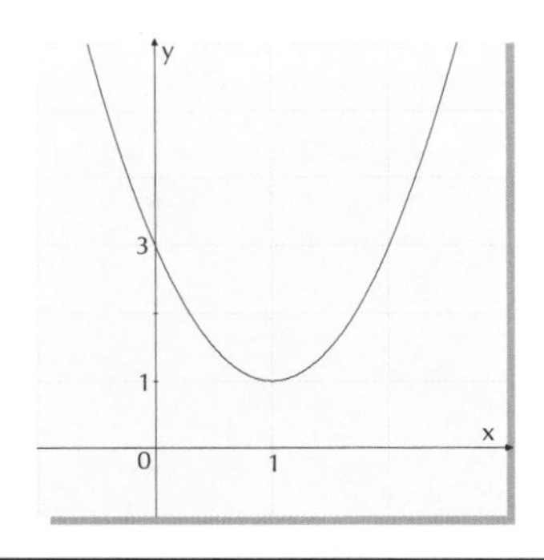

Sketching Quadratic Graphs

A) up or down — decide which direction the curve points in.

B) axes — find where the curve crosses them.

C) max / min — find the turning point.

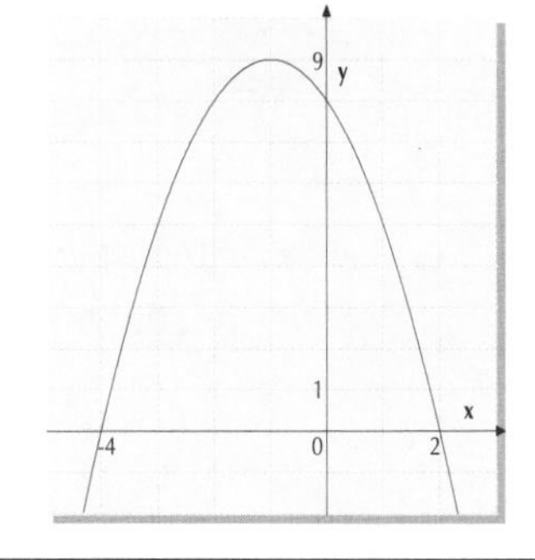

Van Gogh, Monet — all the greats started out sketching graphs...

So there's three steps here to learn. Simple enough. You can do the third step (finding the max/min point) by either a) completing the square, which is covered a bit later, or b) using the fact that the graph's symmetrical — so once you've found the points where it crosses the x-axis, the point halfway between them will be the max/min. It's all laughs here...

Factorising a Quadratic

Factorising a quadratic means putting it into two brackets — and is useful if you're trying to draw a graph of a quadratic or solve a quadratic equation. It's pretty easy if a = 1 (in $ax^2 + bx + c$ form), but can be a real pain otherwise.

$$x^2 - x - 12 = (x-4)(x+3)$$

Factorising's not so bad when a = 1

EXAMPLE: Solve $x^2 - 8 = 2x$ by factorising.

A) Put into $ax^2 + bx + c = 0$ Form

$x^2 - 2x - 8 = 0$ ← So a = 1, b = –2, c = –8.

Write down the two brackets with x's in: $x^2 - 2x - 8 = (x \quad)(x \quad)$

B) Find the Two Numbers

Find two numbers that multiply together to make 'c' but which also add or subtract to give 'b' (you can ignore any minus signs for now).

1 and 8 multiply to give 8 — and add / subtract to give 9 and 7.
2 and 4 multiply to give 8 — and add / subtract to give 6 and 2.

This is the value for 'b' you're after — so this is the right combination: 2 and 4.

C) Find the Signs

Now all you have to do is put in the plus or minus signs.

$x^2 - 2x - 8 = (x \quad 4)(x \quad 2)$

$x^2 - 2x - 8 = (x+2)(x-4)$

If c is negative, then the signs must be different.

It must be +2 and –4 because 2×(–4)=–8 and 2+(–4)=2–4=–2

D) Solve the Equation

All you've done so far is to factorise the equation — you've still got to solve it.

$(x+2)(x-4) = 0$

$\Rightarrow x+2=0$ or $x-4=0$

$\Rightarrow x=-2$ or $x=4$

Don't forget this last step. The factors aren't the answer.

Factorising Quadratics

A) Rearrange the equation into the standard ax^2+bx+c form.

B) Write down the two brackets: (x)(x)

C) Find two numbers that multiply to give 'c' and add / subtract to give 'b' (ignoring signs).

D) Put the numbers in the brackets and choose their signs.

Another Example...

EXAMPLE: Solve $x^2 + 4x - 21 = 0$ by factorising.

This equation is already in the standard format — you can write down the brackets straight away.

$x^2 + 4x - 21 = (x \quad)(x \quad)$

1 and 21 multiply to give 21 — and add / subtract to give 22 and 20.
3 and 7 multiply to give 21 — and add / subtract to give 10 and 4.

This is the value of 'b' you're after — 3 and 7 are the right numbers.

$x^2 + 4x - 21 = (x+7)(x-3)$

And solving the equation to find x gives... $\Rightarrow x=-7$ or $x=3$

Scitardauq Gnisirotcaf — you should know it backwards...

Factorising quadratics — this is very basic stuff. You've really got to be comfortable with it. If you're even slightly rusty, you need to practise it until it's second nature. Remember why you're doing it — you don't factorise simply for the pleasure it gives you — it's so you can solve quadratic equations. Well, that's the theory anyway...

Factorising a Quadratic

It's not over yet...

Factorising a quadratic when $a \neq 1$

These can be a real pain. The basic method's the same as on the previous page — but it can be a bit more awkward.

EXAMPLE: Factorise $3x^2 + 4x - 15$

A) Write Down Two Brackets

As before, write down two brackets — but instead of just having x in each, you need two things that will multiply to give $3x^2$.

$3x^2 + 4x - 15 = (3x \quad)(x \quad)$

It's got to be 3x and x here.

B) The Fiddly Bit

You need to find two numbers that multiply together to make 15 — but which will give you 4x when you multiply them by x and 3x, and then add / subtract them.

$(3x \quad 1)(x \quad 15) \Rightarrow x$ and $45x$ which then add or subtract to give 46x and 44x.

$(3x \quad 15)(x \quad 1) \Rightarrow 15x$ and $3x$ which then add or subtract to give 18x and 12x.

$(3x \quad 3)(x \quad 5) \Rightarrow 3x$ and $15x$ which then add or subtract to give 18x and 12x.

$(3x \quad 5)(x \quad 3) \Rightarrow 5x$ and $9x$ which then add or subtract to give 14x and 4x.

This is the value you're after — so this is the right combination.

C) Add the Signs

You know the brackets must be like these... $(3x \quad 5)(x \quad 3) = 3x^2 + 4x - 15$

'c' is negative — that means the signs in the brackets are different.

So all you have to do is put in the plus or minus signs.

You've only got two choices — if you're unsure, just multiply them out to see which one's right.

$(3x + 5)(x - 3) = 3x^2 - 4x - 15$

or...

$(3x - 5)(x + 3) = 3x^2 + 4x - 15$ ⟸ So it's this one.

Sometimes it's best just to ***Cheat*** *and use the* ***Formula***

Here's two final points to bear in mind:

1) It won't always factorise.
2) Sometimes factorising is so messy that it's easier to just use the quadratic formula...

So if the question doesn't tell you to factorise, don't assume it will factorise.
And if it's something like this thing below, don't bother trying to factorise it...

EXAMPLE: Solve $6x^2 + 87x - 144 = 0$

This will actually factorise, but there's 2 possible bracket forms to try.

$(6x \quad)(x \quad)$ or $(3x \quad)(2x \quad)$ And for each of these, there's 8 possible ways of making 144 to try.

And you can quote me on that...

"He who can properly do quadratic equations is considered a god."
Plato

"Quadratic equations are the music of reason."
James J Sylvester

Completing the Square

Completing the Square is a handy little trick that you should definitely know how to use.
It can be a bit fiddly — but it gives you loads of information about a quadratic really quickly.

Take any old quadratic and put it in a Special Form

Completing the square can be really confusing. For starters, what does "Completing the Square" mean? What is the square? Why does it need completing? Well, there is some logic to it:

1) The square is something like this: $(x + \text{something})^2$ It's basically the factorised equation (with the factors both the same), but there's something missing...

2) ...So you need to 'complete' it by adding a number to the square to make it equal to the original equation. $(x + \text{something})^2 + \text{d}$

You'll start with something like this... ...sort the x-coefficients... ...and you'll end up with something like this.

$2x^2 + 8x - 5$  $2(x+2)^2 + ?$ ⟶ $2(x+2)^2 - 13$

Lovely!

Make completing the square a bit Easier

There are only a few stages to completing the square — if you can't be bothered trying to understand it, just learn how to do it. But I reckon it's worth spending a bit more time to get your head round it properly.

A) Take Out a Factor of 'a' — take a factor of a out of the x^2 and x terms.

$f(x) = 2x^2 + 3x - 5$ ← This is in the form $ax^2 + bx + c$

This '2' is an 'a'.

$f(x) = 2\left(x^2 + \frac{3}{2}x\right) - 5$ ← Check that the bracket multiplies out to what you had before.

This is $\frac{b}{a}$

B) Rewrite the Bracket — rewrite the bracket as one bracket squared.

The number in the brackets is always half the old number in front of the x. $\frac{b}{2a}$

$f(x) = 2\left(x + \frac{3}{4}\right)^2 + \text{d}$ ← d is a number you have to find to make the new form equal to the old one.

Don't forget the 'squared' sign.

C) Complete the Square — find d.

To do this, make the old and new equations equal each other...

$$2\left(x + \frac{3}{4}\right)^2 + \text{d} = 2x^2 + 3x - 5$$

...and you can find d.

$$2x^2 + 3x + \frac{9}{8} + \text{d} = 2x^2 + 3x - 5$$

The x^2 and x bits are the same on both sides so they can disappear.

$$\frac{9}{8} + \text{d} = -5$$

$$\Rightarrow \text{d} = -\frac{49}{8}$$

Completing the Square

A) THE BIT IN THE BRACKETS IS ALWAYS — $a\left(x + \frac{b}{2a}\right)^2$

B) CALL THE NUMBER AT THE END d — $a\left(x + \frac{b}{2a}\right)^2 + \text{d}$

C) MAKE THE TWO FORMS EQUAL — $ax^2 + bx + c = a\left(x + \frac{b}{2a}\right)^2 + \text{d}$

D) 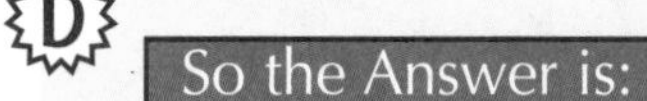

$$f(x) = 2x^2 + 3x - 5 = 2\left(x + \frac{3}{4}\right)^2 - \frac{49}{8}$$

Complete your square — it'd be root not to...

Remember — you're basically trying to write the expression as one bracket squared, but it doesn't quite work. So you have to add a number (d) to make it work. It's a bit confusing at first, but once you've learnt it you won't forget it in a hurry.

Completing the Square

Once you've completed the square, you can very quickly say loads about a quadratic function. And it all relies on the fact that a squared number can never be less than zero... ever.

Completing the square can sometimes be *Useful*

This is a quadratic written as a completed square. As it's a quadratic function and the coefficient of x^2 is positive, it's a u-shaped graph.

$$f(x) = 3x^2 - 6x - 7 = 3(x-1)^2 - 10$$

This is a square — it can never be negative. The smallest it can be is 0.

A **Find the Minimum** — make the bit in the brackets equal to zero.

When the squared bit is zero, f(x) reaches its minimum value. This means the graph reaches its lowest point.

$$f(x) = 3(x-1)^2 - 10$$

This number here is the minimum.

$$f(1) = 3(1-1)^2 - 10$$

$$f(1) = 3(0)^2 - 10 = -10$$

f(1) means using x=1 in the function

So the minimum is -10, when x=1

B **Where Does f(x) Cross the x-axis?** — i.e. find x.

Make the completed square function equal zero.

$$3(x-1)^2 - 10 = 0$$

Solve it to find where f(x) crosses the x-axis.

$$\Rightarrow (x-1)^2 = \frac{10}{3}$$

$$\Rightarrow x - 1 = \pm\sqrt{\frac{10}{3}}$$

da-de-dah ... rearranging again.

$$\Rightarrow x = 1 \pm \sqrt{\frac{10}{3}}$$

So f(x) crosses the x-axis when... $x = 2.83$ or -0.83

These notes are all about graphs with positive coefficients in front of the x^2. But if the coefficient is negative, then the graph is flipped upside-down (n-shaped, not u-shaped).

With this information, you can easily sketch the graph...

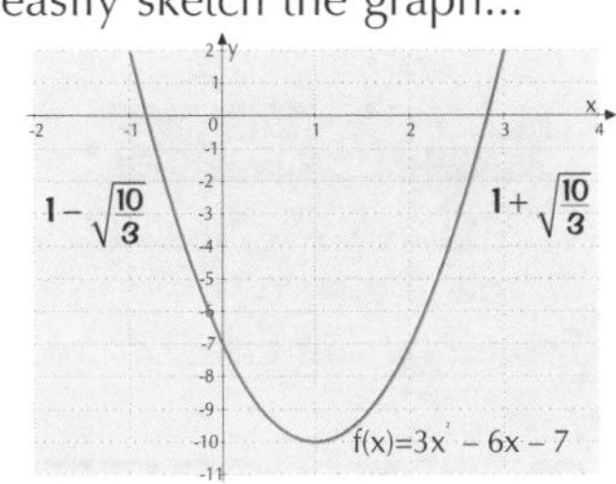

Some functions don't have *Real Roots*

By completing the square, you can also quickly tell if the graph of a quadratic function ever crosses the x-axis. It'll only cross the x-axis if the function changes sign (i.e. goes from positive to negative or vice versa). Take this function...

Find the Roots

$$f(x) = x^2 + 4x + 7$$

$$f(x) = (x+2)^2 + 3$$

This number's positive.

The smallest this bit can be is zero (at x = −2).

$(x + 2)^2$ is never less than zero so f(x) is never less than three.

This means that:

a) f(x) can never be negative.
b) The graph of f(x) never crosses the x-axis.

If the coefficient of x^2 is negative, you can do the same sort of thing to check whether f(x) ever becomes positive.

Don't forget — two wrongs don't make a root...

You'll be pleased to know that that's the end of me trying to tell you how to do something you probably really don't want to do. Now you can push it to one side and run off to roll around in a bed of nettles... much more fun.

The Quadratic Formula

Unlike factorising, the quadratic formula always works... no ifs, no buts, no butts, no nothing...

The **Quadratic Formula** — a reason to be cheerful, but careful...

If you want to solve a quadratic equation $ax^2 + bx + c = 0$, then the answers are given by this formula:

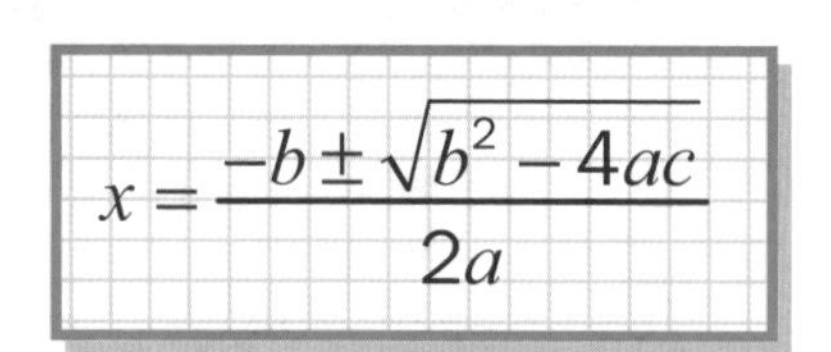

$$x = \frac{-b \pm \sqrt{b^2 - 4ac}}{2a}$$

The formula's a godsend — but use the power wisely...

If any of the coefficients (i.e. if a, b or c) in your quadratic equation are negative — be especially careful.

Always take things nice and slowly — don't try to rush it.

It's a good idea to write down what a, b and c are before you start plugging them into the formula.

There are a couple of minus signs in the formula — which can catch you out if you're not paying attention.

I shall teach you the ways of the **Formula**

EXAMPLE: Solve the quadratic equation $3x^2 - 4x = 8$ to 2 d.p.

The mention of decimal places is a big clue that you should use the formula.

Rearrange the Equation

Get the equation in the standard $ax^2 + bx + c = 0$ form.

$3x^2 - 4x = 8$

$3x^2 - 4x - 8 = 0$

Find *a*, *b* and *c*

Write down the coefficients a, b and c — making sure you don't forget minus signs.

$3x^2 - 4x - 8 = 0$

$a = 3 \quad b = -4 \quad c = -8$

Stick Them in the Formula

Very carefully, plug these numbers into the formula. It's best to write down each stage as you do it.

$$x = \frac{-b \pm \sqrt{b^2 - 4ac}}{2a}$$

$$= \frac{-(-4) \pm \sqrt{(-4)^2 - 4 \times 3 \times (-8)}}{2 \times 3}$$

$$= \frac{4 \pm \sqrt{16 + 96}}{6}$$

$$= \frac{4 \pm \sqrt{112}}{6}$$

$$= \frac{4 \pm 10.5830}{6}$$

$$x = 2.43 \ or -1.10$$

Until you've finished, keep at least one or two more decimal places than you'll eventually need.

Don't round numbers up or down unless you have to — use surds where possible. (See page 3)

The $\pm$ sign means that we have two different expressions for x — which you get by replacing the $\pm$ with + and −.

Round to 2 d.p. (or whatever you've been asked to round to) but only when you've finished the calculation.

Using this magic formula, I shall take over the world... ha ha ha...

Okay, maybe it's not quite that good... but it's really important. So learn it properly — which means spending enough time until you can just say it out loud the whole way through, with no hesitations. Or perhaps you could try singing it as loud as you can to the tune of your favourite cheesy song. Sha-la-la-la-la-la-la-ha... La-di-da... Sha-la-la-la-la-la-la-ha... La-di-da... Sha-la-la-la-la-la-la-ha...

The Quadratic Formula

By using part of the quadratic formula, you can quickly tell if a quadratic equation has two solutions, one solution, or no solutions at all. Tell me more, I hear you cry...

How Many Roots? Check the $b^2 - 4ac$ bit...

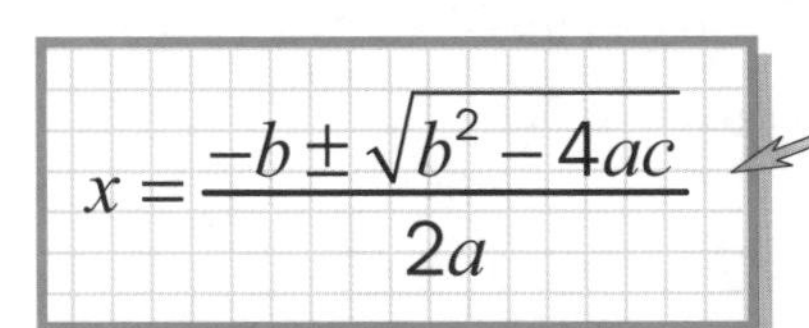

When you try to find the roots of a quadratic function, this bit in the square-root sign ($b^2 - 4ac$) can be positive, zero, or negative. It's this that tells you if a quadratic function has two roots, one root, or no roots.

The $b^2 - 4ac$ bit is called the discriminant.

Because — if the discriminant is positive, the formula will give you two different values — when you add or subtract the $\sqrt{b^2 - 4ac}$ bit.

But if it's zero, you'll only get one value, since adding or subtracting zero doesn't make any difference.

And if it's negative, you don't get any (real) values because you can't take the square root of a negative number.

Well, not in Core 1. In later modules, you can actually take the square root of negative numbers and get 'imaginary' numbers. That's why we say no 'real' roots — because there are 'imaginary' roots!

It's good to be able to picture what this means:

A root is just when y = 0, so it's where the graph touches or crosses the x-axis.

$b^2 - 4ac > 0$
Two roots

So the graph crosses the x-axis twice and these are the roots:

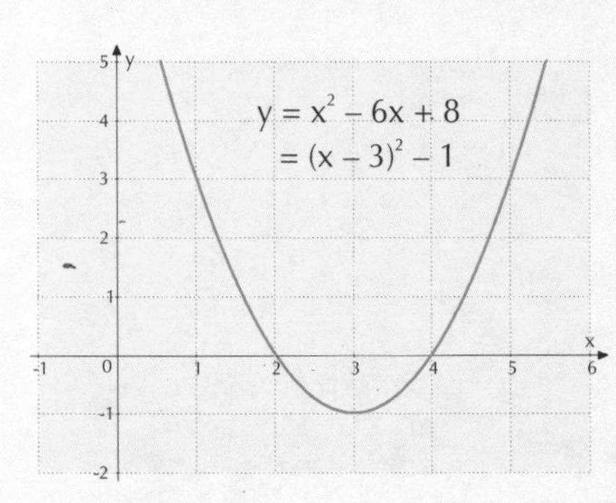

$b^2 - 4ac = 0$
One root

The graph just touches the x-axis from above (or from below if the x^2 coefficient is negative).

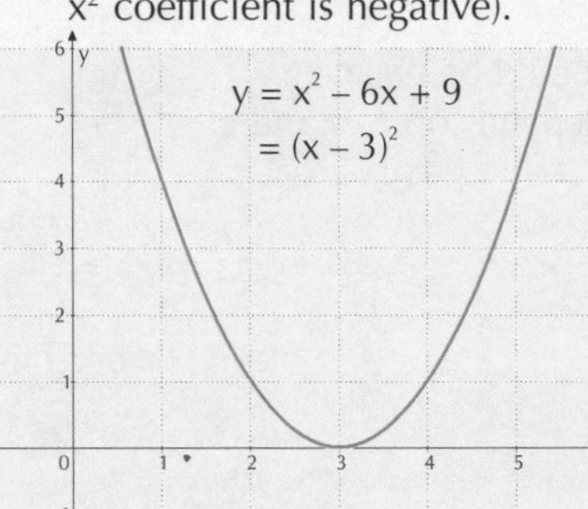

$b^2 - 4ac < 0$
No roots

The graph doesn't touch the x-axis at all.

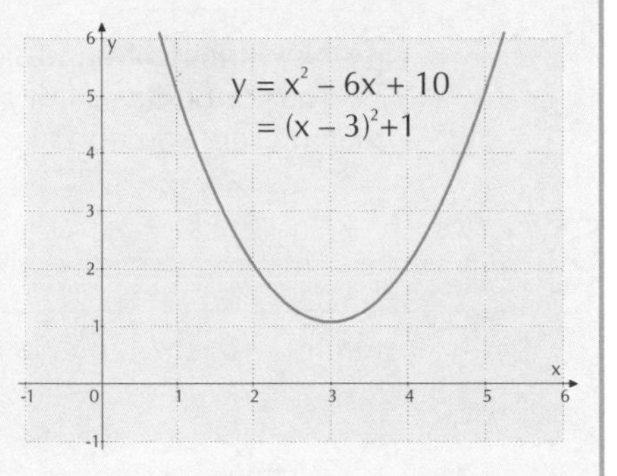

EXAMPLE: Find the range of values of k for which: a) f(x)=0 has 2 distinct roots, b) f(x)=0 has 1 root, c) f(x) has no real roots, where $f(x) = 3x^2 + 2x + k$.

First of all, work out what the discriminant is: $b^2 - 4ac = 2^2 - 4 \times 3 \times k = 4 - 12k$

These calculations are exactly the same. You don't need to do them if you've done a) because the only difference is the equality symbol.

a) Two distinct roots means:
$$b^2 - 4ac > 0 \Rightarrow 4 - 12k > 0 \Rightarrow 4 > 12k \Rightarrow k < \tfrac{1}{3}$$

b) One root means:
$$b^2 - 4ac = 0 \Rightarrow 4 - 12k = 0 \Rightarrow 4 = 12k \Rightarrow k = \tfrac{1}{3}$$

c) No roots means:
$$b^2 - 4ac < 0 \Rightarrow 4 - 12k < 0 \Rightarrow 4 < 12k \Rightarrow k > \tfrac{1}{3}$$

ha ha ha ha haaaaaa... ha ha ha... ha ha ha ... ha ha ha........

So for questions about "how many roots", think discriminant — i.e. $b^2 - 4ac$. And don't get the inequality signs (> and <) the wrong way round. It's obvious, if you think about it.

Essential Proofs

All the stuff you know about the quadratic formula can be proved. And you've got to be able to prove it. Remember you've got to get the equations in the form $ax^2 + bx + c = 0$ before you do anything.

Prove the quadratic formula by Completing the Square

The quadratic formula is a wonderful thing. And don't just accept my word that the formula's true — prove it.

Say you've got a quadratic equation in standard form... $ax^2+bx+c=0$ ...and you need to find x.

① The first thing to do is complete the square.

$$ax^2+bx+c=0$$

$$a\left(x^2+\frac{b}{a}x\right)+c=0$$

$$a\left(x+\frac{b}{2a}\right)^2-\frac{b^2}{4a}+c=0$$

Take a common factor of 'a' out of the first two terms — then it's easier to see how to complete the square.

What you're trying to do is find x — get it on its own on one side.

② Then, rearrange the formula. This is much better, because there's only one x now — so you can find what x actually is.

$$a\left(x+\frac{b}{2a}\right)^2=\frac{b^2}{4a}-c=\frac{b^2-4ac}{4a}$$

$$\left(x+\frac{b}{2a}\right)^2=\frac{b^2-4ac}{4a^2}$$

Divide both sides by a.

Take the last two terms over to the right-hand side, and then add them together as fractions.

③ Now the right-hand side could be negative, zero or positive — it all depends on a, b and c.

$$x+\frac{b}{2a}=\pm\sqrt{\frac{b^2-4ac}{4a^2}}=\pm\frac{\sqrt{b^2-4ac}}{2a}$$

$$x=-\frac{b}{2a}\pm\frac{\sqrt{b^2-4ac}}{2a}$$

$$x=\frac{-b\pm\sqrt{b^2-4ac}}{2a}$$

And this is the quadratic formula that you've come to know and love.

If you're good enough at algebra, you might be able to get away with not learning all this proof, and doing all the rearranging yourself.

Just remember:

You start with $ax^2+bx+c=0$

And end with $x=\frac{-b\pm\sqrt{b^2-4ac}}{2a}$

Forget maths, do French instead...

The stuff on this page can really only be used one way — and that's by learning it. Not what you want to hear, but it's the truth. And the truth hurts sometimes, don't it? Well listen buster, you've just got to get your head down, and get on with it — no slacking. Unless you've already passed AS Maths, and you're just reading this for fun...

Cows

The stuff on this page isn't strictly on the syllabus. But I've included it anyway because I reckon it's really important stuff that you ought to know.

There are loads of Different Types of Cows

Dairy Cattle

Every day a dairy cow can produce up to 128 pints of milk — which can be used to make 14 lbs of cheese, 5 gallons of ice cream, or 6 lbs butter.

The Jersey

The Jersey is a small breed best suited to pastures in high rainfall areas. It is kept for its creamy milk.

Advantages
1) *Can produce creamy milk until old age.*
2) *Milk is the highest in fat of any dairy breed (5.2%).*
3) *Fairly docile, although bulls can't be trusted.*

Disadvantages
1) *Produces less milk than most other breeds.*

The Holstein-Friesian

This breed can be found in many areas.
It is kept mainly for milk.

Advantages
1) *Produce more milk than any breed.*
2) *The breed is large, so bulls can be sold for beef.*

Disadvantages
1) *Milk is low in fat (3.5%).*

Beef Cattle

Cows are sedentary animals who spend up to 8 hours a day chewing the cud while standing still or lying down to rest after grazing. Getting fat for people to eat.

The Angus

The Angus is best suited to areas where there is moderately high rainfall.

Advantages
1) *Early maturing.*
2) *High ratio of meat to body weight.*
3) *Forages well.*
4) *Adaptable.*

The Hereford

The Hereford matures fairly early, but later than most shorthorn breeds. All Herefords have white faces, and if a Hereford is crossbred with any other breed of cow, all the offspring will have white or partially white faces.

Advantages
1) *Hardy.*
2) *Adaptable to different feeds.*

Disadvantages
1) *Susceptible to eye diseases.*

Milk comes from Cows

This is really important — try not to forget it.

Milk is an emulsion of butterfat suspended in a solution of water (roughly 80%), lactose, proteins and salts. Cow's milk has a specific gravity around 1.03.
It's pasteurised by heating it to 63° C for 30 minutes. It's then rapidly cooled and stored below 10° C .

Louis Pasteur began his experiments into 'pasteurisation' in 1856.
By 1946, the vacuum pasteurisation method had been perfected, and in 1948, UHT (ultra heat-treated) pasteurisation was introduced.

cow + grass = fat cow

fat cow + milking machine ⇒ milk

You will often see cows with pieces of grass sticking out of their mouths.

SOME IMPORTANT FACTS TO REMEMBER:
- A newborn calf can walk on its own an hour after birth
- A cow's teeth are only on the bottom of her mouth
- While some cows can live up to 40 years, they generally don't live beyond 20.

Cows on the Internet

For more information on cows, try these websites:

www.allcows.com (including Cow of the Month)
www.crazyforcows.com (with cow e-postcards)
www.moomilk.com (includes a 'What's the cow thinking?' contest.)
http://www.geocities.com/Hollywood/9317/meowcow.html
(for cow-tipping on the Internet)

The Cow
The cow is of the bovine ilk;
One end is moo, the other, milk.

— Ogden Nash

Famous Cows and Cow Songs

Famous Cows
1) Ermintrude from the Magic Roundabout.
2) Graham Heifer — the Boddingtons cow.
3) Other TV commercial cows — Anchor, Dairylea
4) The cow that jumped over the moon.
5) Greek Mythology was full of gods turning themselves and their girlfriends into cattle.

Cows in Pop Music
1) Size of a Cow — the Wonder Stuff
2) Saturday Night at the Moo-vies — The Drifters
3) What can I do to make you milk me? — The Cows
4) One to an-udder — the Charlatans
5) Milk me baby, one more time — Britney Spears

Where's me Jersey — I'm Friesian...

Cow-milking — an underrated skill, in my opinion. As Shakespeare once wrote, 'Those who can milk cows are likely to get pretty good grades in maths exams, no word of a lie'. Well, he probably would've written something like that if he was into cows. And he would've written it because cows are helpful when you're trying to work out what a question's all about — and once you know that, you can decide the best way forward. And if you don't believe me, remember the saying of the ancient Roman Emperor Julius Caesar, 'If in doubt, draw a cow'.

Factorising Cubics

Factorising a quadratic function is okay — but you might also be asked to factorise a cubic (something with x^3 in it). And that takes a bit more time — there are more steps, so there are more chances to make mistakes.

Factorising a cubic given One Factor

$$f(x) = 2x^3 + x^2 - 8x - 4$$

Factorising a cubic means exactly what it meant with a quadratic — putting brackets in.
When they ask you to factorise a cubic equation, they'll usually tell you one of the factors.

EXAMPLE: Given that $(x + 2)$ is a factor of $f(x) = 2x^3 + x^2 - 8x - 4$, express $f(x)$ as the product of three linear factors.

① The first step is to find a quadratic factor. So write down the factor you know, along with another set of brackets.

$$(x+2)(\qquad\qquad) = 2x^3 + x^2 - 8x - 4$$

Put the x^2 bit in this new set of brackets.
These have to multiply together to give you this.

$$(x+2)(2x^2 \qquad) = 2x^3 + x^2 - 8x - 4$$

② Find the number for the second set of brackets.
These have to multiply together to give you this.

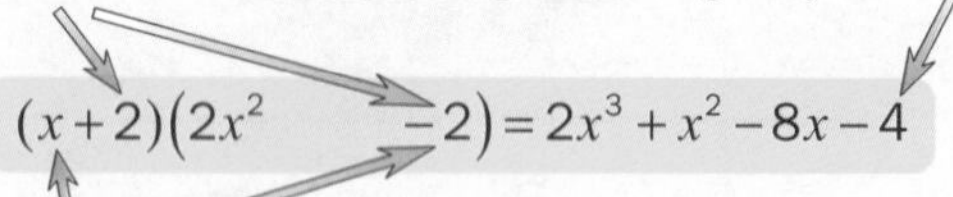

$$(x+2)(2x^2 \qquad -2) = 2x^3 + x^2 - 8x - 4$$

③ These multiplied give you $-2x$, but there's $-8x$ in $f(x)$ — so you need an 'extra' $-6x$. And that's what this $-3x$ is for.

$$(x+2)(2x^2 - 3x - 2) = 2x^3 + x^2 - 8x - 4$$

You only need $-3x$ because it's going to be multiplied by 2 — which makes $-6x$.

Factorising Cubics

1) **Write down the factor $(x - k)$.**
2) **Put in the x^2 term.**
3) **Put in the constant.**
4) **Put in the x term by comparing the number of x's on both sides.**
5) **Check there are the same number of x^2's on both sides.**
6) **Factorise the quadratic you've found — if that's possible.**

④ Before you go any further, check that there are the same number of x^2's on both sides.

$4x^2$ from here...

$$(x+2)(2x^2 - 3x - 2) = 2x^3 + x^2 - 8x - 4$$

...and $-3x^2$ from here... ...add together to give this x^2.

If this is okay, factorise the quadratic into two linear factors.

$$(2x^2 - 3x - 2) = (2x + 1)(x - 2)$$

And so... $2x^3 + x^2 - 8x - 4 = (x+2)(2x+1)(x-2)$

If you wanted to solve a cubic, you'd do it exactly the same way — put it in the form $ax^3 + bx^2 + cx + d = 0$ and factorise.

Factorising a cubic given No Factors

If they don't give you the first factor, you have to find it yourself. But it's okay — they'll give you an easy one.
The best way to find a factor is to guess — use trial and error.

Find f(1) If the answer is zero, you know $(x - 1)$ is a factor.
If the answer isn't zero, find $f(-1)$. If that's zero, then $(x + 1)$ is a factor.

If that doesn't work, keep trying small numbers (find $f(2)$, $f(-2)$, $f(3)$, $f(-3)$ and so on) until you find a number that gives you zero when you put it in the cubic. Call that number k.

$(x - k)$ is a factor of the cubic (from the Factor Theorem).

I love the smell of fresh factorised cubics in the morning...

Factorising cubics is exactly the same as learning to unicycle... It's impossible at first. But when you finally manage it, it's really easy from then onwards and you'll never forget it. Probably. To tell the truth, I can't unicycle at all. So don't believe a word I say.

Section Two Revision Questions

Mmmm, well, quadratic equations — not exactly designed to make you fall out of your chair through laughing so hard, are they? But (and that's a huge 'but') they'll get you plenty of marks come that fine morning when you march confidently into the exam hall — if you know what you're doing. And what better way to make sure you know what you're doing than to practise. So here we go then, on the thrill-seekers' ride of a lifetime — the CGP quadratic equation revision section...

1) Factorise the following expressions. While you're doing this, sing a jolly song to show how much you enjoy it.
   a) x^2+2x+1, b) $x^2-13x+30$, c) x^2-4, d) $3+2x-x^2$
   e) $2x^2-7x-4$, f) $5x^2+7x-6$.
2) Solve the following equations. And sing verse two of your jolly song.
   a) $x^2-3x+2=0$, b) $x^2+x-12=0$, c) $2+x-x^2=0$, d) $x^2+x-16=x$
   e) $3x^2-15x-14=4x$, f) $4x^2-1=0$, g) $6x^2-11x+9=2x^2-x+3$.
3) Rewrite these quadratics by completing the square. Then state their maximum or minimum value and the value of x where this occurs. Also, say which ones cross the x-axis — just for a laugh, like.
   a) x^2-4x-3, b) $3-3x-x^2$, c) $2x^2-4x+11$, d) $4x^2-28x+48$.
4) How many roots do these quadratics have? Sketch their graphs.
   a) $x^2-2x-3=0$, b) $x^2-6x+9=0$, c) $2x^2+4x+3=0$.
5) Solve these quadratic equations to two decimal places.
   a) $3x^2-7x+3=0$, b) $2x^2-6x-2=0$, c) $x^2+4x+6=12$.
6) If the quadratic equation $x^2+kx+4=0$ has two roots, what are the possible values of k?
7) For the quadratic equation $ax^2+bx+c=0$ prove that $x=\frac{-b\pm\sqrt{b^2-4ac}}{2a}$

OK, I think that's enough. Go and make yourself a cup of tea. Treat yourself to a chocolate biscuit.

Here is a new way to enjoy Pelican biscuits:

Bite a small piece off two opposite corners of a Pelican.
Immerse one corner in coffee, and suck coffee up through the Pelican, like a straw. You will need to suck quite hard to start with.
After a few seconds you will notice the biscuit part of the Pelican start to lose structural integrity. At this point, cram it into your mouth, where it will collapse into a mass of hot molten chocolate, biscuit and coffee.

Mmmm.

Linear Inequalities

Solving inequalities is very similar to solving equations. You've just got to be really careful that you keep the inequality sign pointing the right way.

> Find the ranges of x that satisfy these inequalities:
> (i) $x-3<-1+2x$ (ii) $8x+2\geq 2x+17$ (iii) $4-3x\leq 16$ (iv) $36x<6x^2$

Sometimes the inequality sign **Changes Direction**

Like I said, these are pretty similar to solving equations — because whatever you do to one side, you have to do to the other. But multiplying or dividing by negative numbers changes the direction of the inequality sign.

Adding *or* ***Subtracting*** *doesn't change the direction of the inequality sign*

EXAMPLE: If you add or subtract something from both sides of an inequality, the inequality sign doesn't change direction.

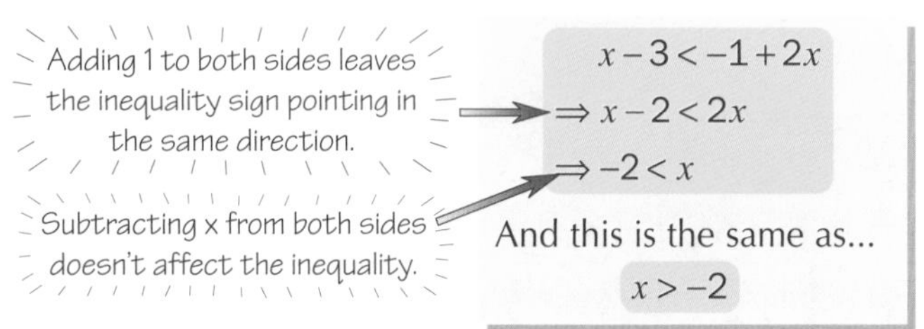

Multiplying *or* ***Dividing*** *by something* ***Positive*** *doesn't affect the inequality sign*

EXAMPLE: Multiplying or dividing both sides of an inequality by a positive number doesn't affect the direction of the inequality sign.

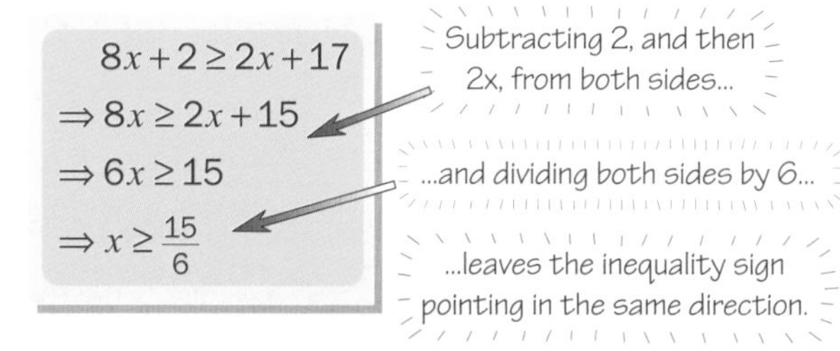

But **Change** *the inequality if you* **Multiply** *or* **Divide** *by something* **Negative**

But multiplying or dividing both sides of an inequality by a negative number changes the direction of the inequality.

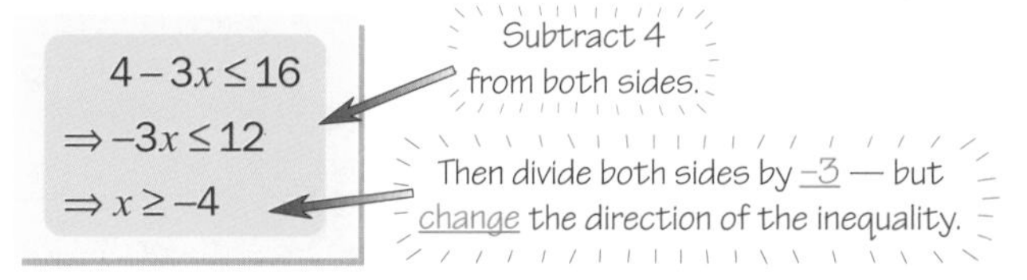

> The reason for the sign changing direction is because it's just the same as swapping everything from one side to the other:
> $-3x\leq 12 \Rightarrow -12\leq 3x \Rightarrow x\geq -4$

Don't *divide both sides by* **Variables** *— like x and y*

You've got to be really careful when you divide by things that might be negative — well basically, don't do it.

EXAMPLE: $36x<6x^2$

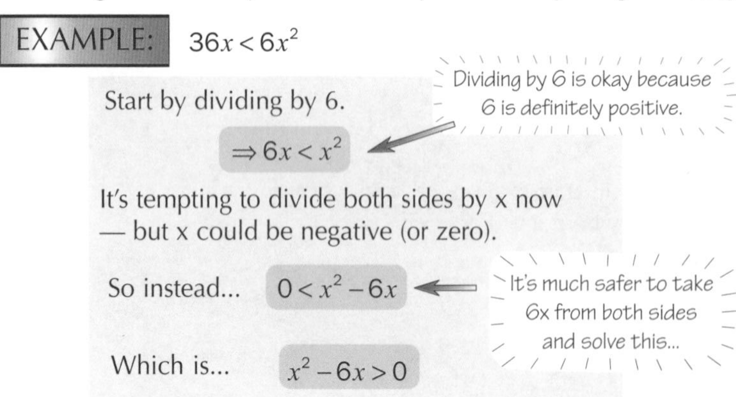

See the next page for more on solving quadratic inequalities.

> ***Two types of inequality sign***
>
> **There are two kinds of inequality sign:**
>
> **Type 1:** $<$ **— less than**
> $>$ **— greater than**
>
> **Type 2:** $\leq$ **— less than or equal to**
> $\geq$ **— greater than or equal to**
>
> **Whatever type the question uses — use the same kind all the way through your answer.**

So no one knows we've arrived safely — splendid...

So just remember — inequalities are just like normal equations except that you have to reverse the sign when multiplying or dividing by a negative number. And don't divide both sides by variables. (You should know not to do this with normal equations anyway because the variable could be zero.) OK — lecture's over.

Quadratic Inequalities

With quadratic inequalities, you're best off drawing the graph and taking it from there.

*Draw a **Graph** to solve a **Quadratic** inequality*

Example: Find the ranges of x which satisfy these inequalities:

① $-x^2+2x+4 \geq 1$

First rewrite the inequality with zero on one side.

$$-x^2+2x+3 \geq 0$$

Then draw the graph of $y=-x^2+2x+3$:

So find where it crosses the x-axis (i.e. where y=0):

$$-x^2+2x+3=0 \Rightarrow x^2-2x-3=0$$
$$\Rightarrow (x+1)(x-3)=0$$
$$\Rightarrow x=-1 \text{ or } x=3$$

And the coefficient of x^2 is negative, so the graph is n-shaped. So it looks like this:

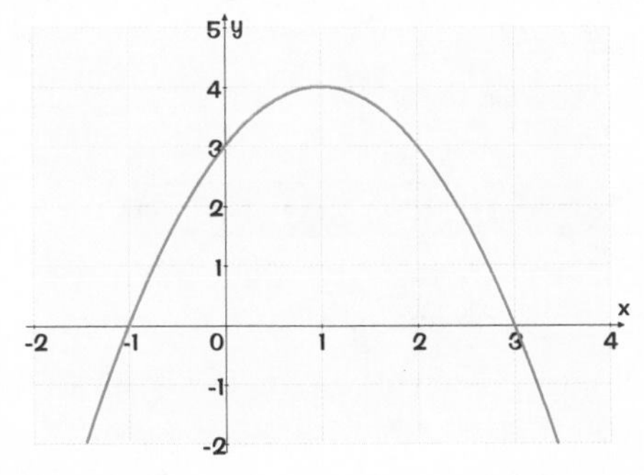

You're interested in when this is positive or zero, i.e. when it's above the x-axis.

From the graph, this is when x is between –1 and 3 (including those points). So your answer is...

$$-x^2+2x+4 \geq 1 \text{ when } -1 \leq x \leq 3.$$

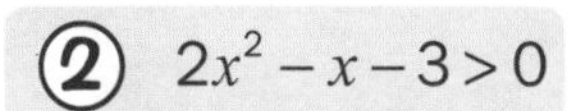

② $2x^2-x-3>0$

This one already has zero on one side, so draw the graph of $y=2x^2-x-3$.

Find where it crosses the x-axis:

$$2x^2-x-3=0$$
$$\Rightarrow (2x-3)(x+1)$$
$$\Rightarrow x=\tfrac{3}{2} \text{ or } x=-1$$

Factorise it to find the roots.

And the coefficient of x^2 is positive, so the graph is u-shaped. And looks like this:

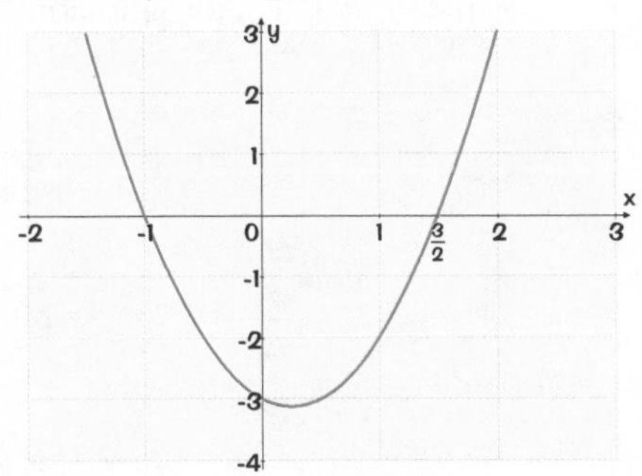

You need to say when this is positive. Looking at the graph, there are two parts of the x-axis where this is true — when x is less than –1 and when x is greater than 3/2. So your answer is:

$$2x^2-x-3>0 \text{ when } x<-1 \text{ or } x>\tfrac{3}{2}.$$

Example (revisited): On the last page you had to solve $36x<6x^2$.

$$36x<6x^2$$
equation 1 ⟹ $\Rightarrow 6x<x^2$
$$\Rightarrow 0<x^2-6x$$

So draw the graph of

$$y=x^2-6x=x(x-6)$$

And this is positive when $x<0$ or $x>6$.

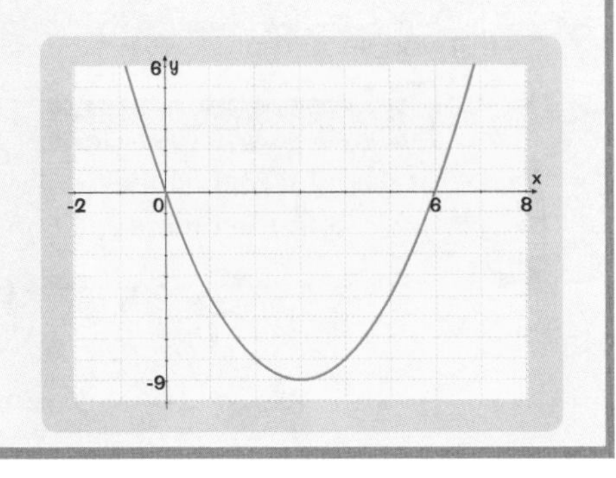

If you divide by x in equation 1, you'd only get half the solution — you'd miss the $x<0$ part.

That's nonsense — I can see perfectly...

Call me sad, but I reckon these questions are pretty cool. They look a lot more difficult than they actually are and you get to draw a picture. Wow! When you do the graph, the important thing is to find where it crosses the x-axis (you don't need to know where it crosses the y-axis) and make sure you draw it the right way up. Then you just need to decide which bit of the graph you want. It'll either be the range(s) of x where the graph is below the x-axis or the range(s) where it's above. And this depends on the inequality sign.

Simultaneous Equations

Solving simultaneous equations means finding the answers to two equations at the same time — i.e. finding values for x and y for which both equations are true. And it's one of those things that you'll have to do again and again — so it's definitely worth practising them until you feel really confident.

① $3x+5y=-4$
② $-2x+3y=9$

This is how simultaneous equations are usually shown. It's a good idea to label them as equation ① and equation ② — so you know which one you're working with.

But they'll look different sometimes, maybe like this. Make sure you rearrange them as 'ax + by = c'.

$4+5y=-3x$
$-2x=9-3y$

rearrange as ax + by = c

$3x+5y=-4$
$-2x+3y=9$

Solving them by Elimination

Elimination is a lovely method. It's really quick when you get the hang of it — you'll be doing virtually all of it in your head.

EXAMPLE:

① $3x+5y=-4$
② $-2x+3y=9$

A Match the Coefficients

Multiply the equations by numbers that will make either the x's or the y's match in the two equations. (Ignoring minus signs.)

Go for the lowest common multiple (LCM). e.g. LCM of 2 and 3 is 6.

To get the x's to match, you need to multiply the first equation by 2 and the second by 3:

①×2 $6x+10y=-8$
②×3 $-6x+9y=27$

B Eliminate to Find One Variable

If the coefficients are the same sign, you'll need to subtract one equation from the other.

If the coefficients are different signs, you need to add the equations.

Add the equations together to eliminate the x's.

C Find the Variable You Eliminated

When you've found one variable, put its value into one of the original equations so you can find the other variable.

So y is 1. Now stick that value for y into one of the equations to find x:

$y=1$ in ① $\Rightarrow 3x+5=-4$

$3x=-9$

$x=-3$

So the solution is $x=-3$, $y=1$.

But you should always...

D Check Your Answer

...by putting these values into the other equation.

② $-2x+3y=9$
$x=-3$
$y=1$

$-2\times(-3)+3\times1=6+3=9$

If these two numbers are the same, then the values you've got for the variables are right.

Elimination Method

1) Match the coefficients
2) Eliminate and then solve for one variable
3) Find the other variable (that you eliminated)
4) Check your answer

Eliminate your social life — do AS-level maths

This is a fairly basic method that won't be new to you. So make sure you know it. The only possibly tricky bit is matching the coefficients — work out the lowest common multiple of the coefficients of x, say, then multiply the equations to get this number in front of each x.

Simultaneous Equations with Quadratics

Elimination is great for simple equations. But it won't always work. Sometimes one of the equations has not just x's and y's in it — but bits with x^2 and y^2 as well. When this happens, you can only use the substitution method.

*Use Substitution if one equation is **Quadratic***

EXAMPLE: $-x+2y=5$ — (L) ← The linear equation — with only x's and y's in.

$x^2+y^2=25$ — (Q) ← The quadratic equation — with some x^2 and y^2 bits in.

Rearrange the linear equation so that either x or y is on its own on one side of the equals sign.

$$\text{(L)}\quad -x+2y=5$$
$$\Rightarrow x=2y-5$$

Substitute this expression into the quadratic equation...

$$\text{Sub into (Q):}\quad x^2+y^2=25$$
$$\Rightarrow (2y-5)^2+y^2=25$$

...and then rearrange this into the form $ax^2+bx+c=0$, so you can solve it — either by factorising or using the quadratic formula.

$$\Rightarrow \left(4y^2-20y+25\right)+y^2=25$$
$$\Rightarrow 5y^2-20y=0$$
$$\Rightarrow 5y(y-4)=0$$
$$\Rightarrow y=0 \text{ or } y=4$$

One Quadratic and One Linear Eqn

1) **Isolate variable in linear equation**
   Rearrange the linear equation to get either x or y on its own.
2) **Substitute into quadratic equation**
   — to get a quadratic equation in just one variable.
3) **Solve to get values for one variable**
   — either by factorising or using the quadratic formula.
4) **Stick these values in the linear equation**
   — to find corresponding values for the other variable.

Finally put both these values back into the linear equation to find corresponding values for x:

When y = 0: $-x+2y=5$ (L)
$$\Rightarrow x=-5$$

When y = 4: $-x+2y=5$ (L)
$$\Rightarrow -x+8=5$$
$$\Rightarrow x=3$$

So the solutions to the simultaneous equations are: $x=-5$, $y=0$ and $x=3$, $y=4$.

As usual, check your answers by putting these values back into the original equations.

Check Your Answers

x = -5, y = 0: $-(-5)+2\times0=5$ ✓
$(-5)^2+0^2=25$ ✓

x = 3, y = 4: $-(3)+2\times4=5$ ✓
$3^2+4^2=25$ ✓

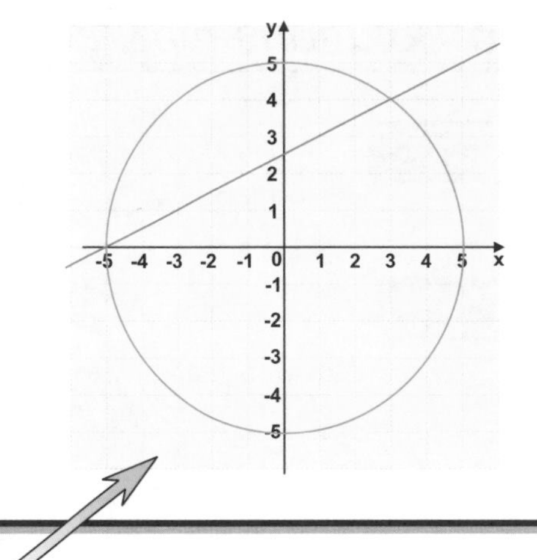

$y = x^2$ — a match-winning substitution...

The quadratic equation above is actually a circle about the origin with radius 5. (Don't worry, you don't need to know about circles till Core 2). The linear equation is just a standard straight line. So what you're actually finding here are the two points where the line passes through the circle. And these turn out to be (–5,0) and (3,4). See the graph. (I thought you might appreciate seeing a graph that wasn't a line or a parabola for a change.)

Geometric Interpretation

When you have to interpret something geometrically — you have to draw a picture and 'say what you see'.

Two Solutions — Two points of Intersection

Example:

$y = x^2 - 4x + 5$ — ①

$y = 2x - 3$ — ②

Solution: Substitute expression for y from ② into ①: $2x - 3 = x^2 - 4x + 5$

Rearrange and solve:
$$x^2 - 6x + 8 = 0$$
$$(x-2)(x-4) = 0$$
$$x = 2 \text{ or } x = 4$$

In ② gives:
$$x = 2 \Rightarrow y = 2\times2 - 3 = 1$$
$$x = 4 \Rightarrow y = 2\times4 - 3 = 5$$

There's 2 pairs of solutions: x=2, y=1 and x=4, y=5

Geometric Interpretation:

So from solving the simultaneous equations, you know that the graphs meet in two places — the points (2,1) and (4,5).

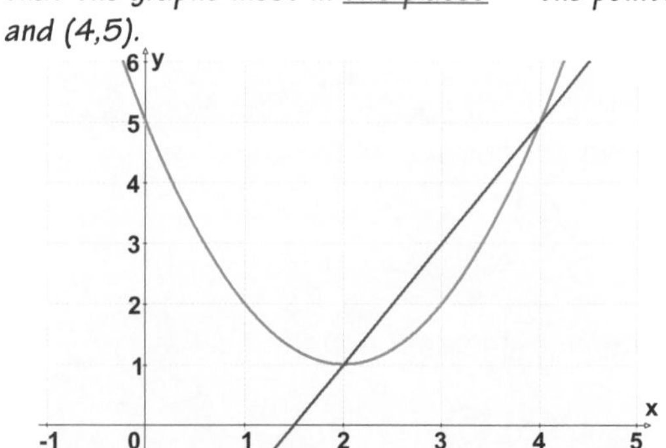

One Solution — One point of Intersection

Example:

$y = x^2 - 4x + 5$ — ①

$y = 2x - 4$ — ②

Solution: Substitute ② in ①: $2x - 4 = x^2 - 4x + 5$

Rearrange and solve:
$$x^2 - 6x + 9 = 0$$
$$(x-3)^2 = 0$$
$$x = 3$$

Double root i.e. you only get 1 solution from the quadratic.

In Equation ② gives:
$$y = 2\times3 - 4$$
$$y = 2$$

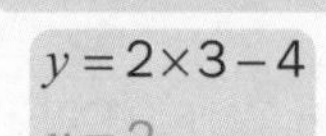

There's 1 solution: x=3, y=2

Geometric Interpretation:

Since the equations have only one solution, the two graphs only meet at one point — (3,2). The straight line is a tangent to the curve.

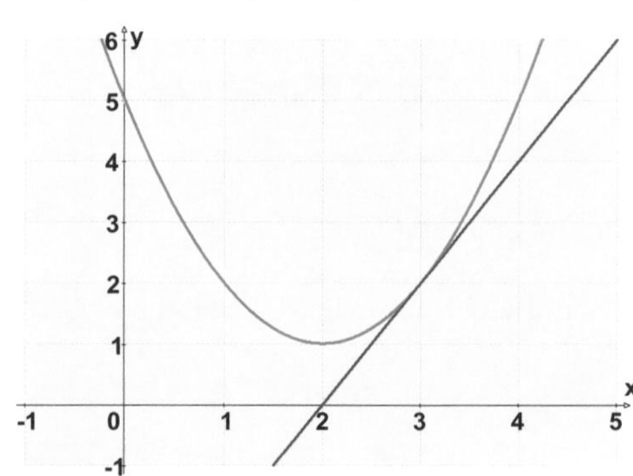

No Solutions means the Graphs Never Meet

Example:

$y = x^2 - 4x + 5$ — ①

$y = 2x - 5$ — ②

Solution: Substitute ② in ①: $2x - 5 = x^2 - 4x + 5$

Rearrange and try to solve with the quadratic formula: $x^2 - 6x + 10 = 0$

$$b^2 - 4ac = (-6)^2 - 4.10$$
$$= 36 - 40 = -4$$

$b^2 - 4ac < 0$, so the quadratic has no roots.

So the simultaneous equations have no solutions.

Geometric Interpretation:

The equations have no solutions — the graphs never meet.

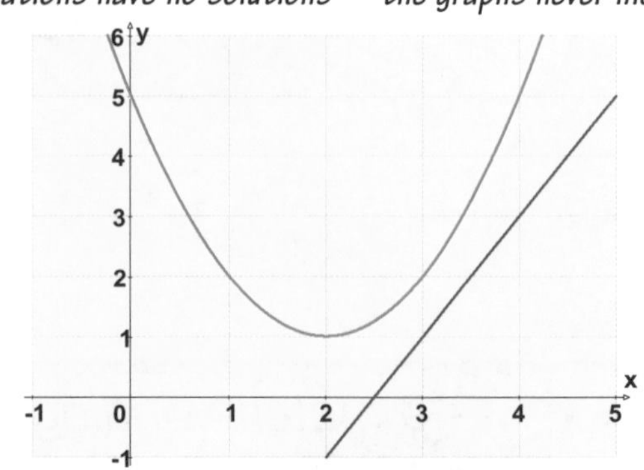

Geometric Interpretation? Frankly my dear, I don't give a damn...

There's some lovely practice Simultaneous Equations questions on the next page.

Section Three Revision Questions

What's that I hear you cry? You want revision questions — and lots of them. Well, it just so happens I've got a few here. Lots of questions on the different kinds of inequalities you need to know about, plus simultaneous equations. Now, as far as quadratic inequalities go, my advice is, 'if you're not sure, draw a picture — even if it's not accurate'. And as for simultaneous equations — well, just don't rush them — or you'll spend twice as long looking for ~~milkshakes~~ mistakes as it took you to do the question in the first place. That's it, the advice is over. So on with the questions...

1) Solve a) $7x - 4 > 2x - 42$, b) $12y - 3 \leq 4y + 4$, c) $9y - 4 \geq 17y + 2$.

2) Find the ranges of x that satisfy these inequalities: i) $x + 6 < 5x - 4$ ii) $4x - 2 > x - 14$ iii) $7 - x \leq 4 - 2x$

3) Find the ranges of x that satisfy the following inequalities. (And watch that you use the right kind of inequality sign in your answers.)

a) $3x^2 - 5x - 2 \leq 0$, b) $x^2 + 2x + 7 > 4x + 9$, c) $3x^2 + 7x + 4 \geq 2(x^2 + x - 1)$.

4) Find the ranges of x that satisfy these jokers: i) $x^2 + 3x - 1 \geq x + 2$ ii) $2x^2 > x + 1$ iii) $3x^2 - 12 < x^2 - 2x$

5) Solve these sets of simultaneous equations.

a) $3x - 4y = 7$ and $-2x + 7y = -22$ b) $2x - 3y = \frac{11}{12}$ and $x + y = -\frac{7}{12}$

6) Find where possible (and that's a bit of a clue) the solutions to these sets of simultaneous equations. Interpret your answers geometrically.

a) $y = x^2 - 7x + 4$, $2x - y - 10 = 0$ b) $y = 30 - 6x + 2x^2$, $y = 2(x + 11)$ c) $x^2 + 2y^2 - 3 = 0$, $y = 2x + 4$

7) A bit trickier: find where the following lines meet:

a) $y = 3x - 4$ and $y = 7x - 5$,
b) $y = 13 - 2x$ and $7x - y - 23 = 0$,
c) $2x - 3y + 4 = 0$ and $x - 2y + 1 = 0$.

Coordinate Geometry

Welcome to geometry club... nice — today I shall be mostly talking about straight lines...

Finding the equation of a line Through Two Points

If you get through your exam without having to find the equation of a line through two points, I'm a Dutchman.

EXAMPLE: Find the equation of the line that passes through the points (–3, 10) and (1, 4), and write it in the forms:

$y - y_1 = m(x - x_1)$ $y = mx + c$ $ax + by + c = 0$

— where a, b and c are integers.

You might be asked to write the equation of a line in any of these forms — but they're all similar.
Basically, if you find an equation in one form — you can easily convert it into either of the others.

The Easiest to find is $y - y_1 = m(x - x_1)$...

Point 1 is (–3, 10) and Point 2 is (1, 4)

Label the Points Label Point 1 as (x_1, y_1) and Point 2 as (x_2, y_2).

Point 1 — $(x_1, y_1) = (-3, 10)$

Point 2 — $(x_2, y_2) = (1, 4)$

It doesn't matter which way round you label them.

Find the Gradient Find the gradient of the line m — this is $m = \frac{y_2 - y_1}{x_2 - x_1}$.

$$m = \frac{4-10}{1-(-3)} = \frac{-6}{4} = -\frac{3}{2}$$

Write Down the Equation Write down the equation of the line, using the coordinates x_1 and y_1 — this is just $y - y_1 = m(x - x_1)$.

$x_1 = -3$ and $y_1 = 10$ ⟹

$$y - 10 = -\frac{3}{2}(x - (-3))$$
$$y - 10 = -\frac{3}{2}(x + 3)$$

...and Rearrange this to get the other two forms:

For the form $y = mx + c$, take everything except the y over to the right.

$$y - 10 = -\frac{3}{2}(x + 3)$$
$$\Rightarrow y = -\frac{3}{2}x - \frac{9}{2} + 10$$
$$\Rightarrow y = -\frac{3}{2}x + \frac{11}{2}$$

To find the form $ax + by + c = 0$, take everything over to one side — and then get rid of any fractions.

Multiply the whole equation by 2 to get rid of the 2's on the bottom line.

$$y = -\frac{3}{2}x + \frac{11}{2}$$
$$\Rightarrow \frac{3}{2}x + y - \frac{11}{2} = 0$$
$$\Rightarrow 3x + 2y - 11 = 0$$

Equations of Lines

1) **LABEL the points (x_1, y_1) and (x_2, y_2).**
2) **GRADIENT — find it and call it m.**
3) **WRITE DOWN THE EQUATION using $y - y_1 = m(x - x_1)$**
4) **CONVERT to one of the other forms, if necessary.**

If you end up with an equation like $\frac{3}{2}x - \frac{4}{3}y + 6 = 0$, where you've got a 2 and a 3 on the bottom of the fractions — multiply everything by the lowest common multiple of 2 and 3, i.e. 6.

There ain't nuffink to this geometry lark, Mister...

This is the sort of stuff that looks hard but is actually pretty easy. Finding the equation of a line in that first form really is a piece of cake — the only thing you have to be careful of is when a point has a negative coordinate (or two). In that case, you've just got to make sure you do the subtractions properly when you work out the gradient. See, this stuff ain't so bad...

Coordinate Geometry

This page is based around two really important facts that you've got to know — one about parallel lines, one about perpendicular lines. It's really a page of unparalleled excitement...

Two more lines...

Line l_1: $3x-4y-7=0$, $y=\frac{3}{4}x-\frac{7}{4}$

Line l_2: $x-3y-3=0$, $y=\frac{1}{3}x-1$

...and two points...

Point A $(3,-1)$

Point B $(-2,4)$

Parallel lines have equal *Gradient*

That's what makes them parallel — the fact that the gradients are the same.

Example: Find the line parallel to l_1 that passes through the point A (3, –1).

Parallel lines have the same gradient.

The original equation is this: $y=\frac{3}{4}x-\frac{7}{4}$

So the new equation will be this: $y=\frac{3}{4}x+c$

We know that the line passes through A, so at this point x will be 3, and y will be –1.

We just need to find c.

Stick these values into the equation to find c.

$$-1=\frac{3}{4}\times 3+c$$

$$\Rightarrow c=-1-\frac{9}{4}=-\frac{13}{4}$$

So the equation of the line is... $y=\frac{3}{4}x-\frac{13}{4}$

And if you're only given the ax + by + c = 0 form it's even easier:

The original line is: $3x-4y-7=0$

So the new line is: $3x-4y-k=0$

Then just use the values of x and y at the point A to find k...

$$3\times 3-4\times(-1)-k=0$$

$$\Rightarrow 13-k=0$$

$$\Rightarrow k=13$$

So the equation is: $3x-4y-13=0$

The gradient of a *Perpendicular* line is: –1 ÷ the Other Gradient

Finding perpendicular lines (or 'normals') is just as easy as finding parallel lines — as long as you remember the gradient of the perpendicular line is –1 ÷ the gradient of the other one.

Example: Find the line perpendicular to l_2 that passes through the point B (–2, 4).

l_2 has equation: $y=\frac{1}{3}x-1$

So if the equation of the new line is y = mx + c, then

$$m=-1\div\frac{1}{3}$$

$$\Rightarrow m=-3$$

Since the gradient of a perpendicular line is: –1 ÷ the other one.

Also...

$$4=(-3)\times(-2)+c$$

$$\Rightarrow c=4-6=-2$$

Putting the coordinates of B(–2, 4) into y = mx + c.

So the equation of the line is... $y=-3x-2$

Or if you start with: l_2 $x-3y-3=0$

To find a perpendicular line, swap these two numbers around, and change the sign of one of them. (So here, 1 and –3 become 3 and 1.)

So the new line has equation... $3x+y+d=0$

Or you could have used –3x – y + d = 0.

But...

$$3\times(-2)+4+d=0$$

$$\Rightarrow d=2$$

Using the coordinates of point B.

And so the equation of the perpendicular line is... $3x+y+2=0$

Wowzers — parallel lines on the same graph dimension...

This looks more complicated than it actually is, all this tangent and normal business. All you're doing is finding the equation of a straight line through a certain point — the only added complication is that you have to find the gradient first. And there's another way to remember how to find the gradient of a normal — just remember that the gradients of perpendicular lines multiply together to make –1.

Curve Sketching

A picture speaks a thousand words... and graphs are what pass for pictures in maths. They're dead useful in getting your head round tricky questions, and time spent learning how to sketch graphs is time well spent.

The graph of $y = kx^n$ is a different shape for different k and n

Usually, you only need a rough sketch of a graph — so just knowing the basic shapes of these graphs will do.

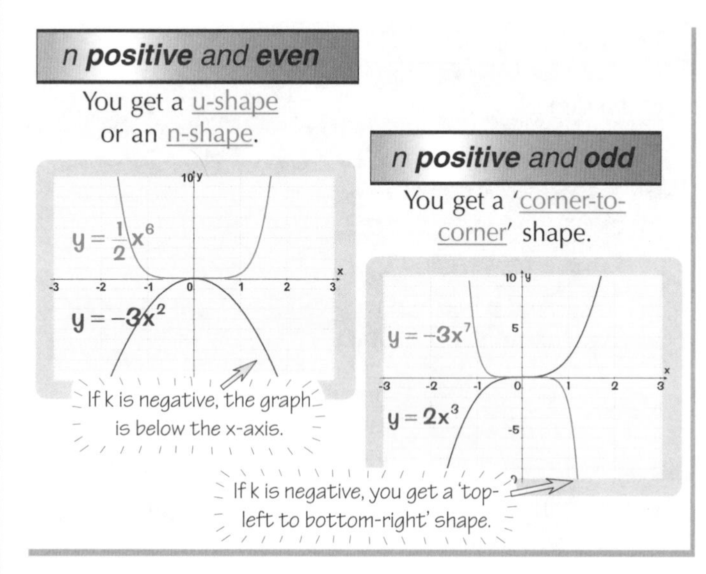

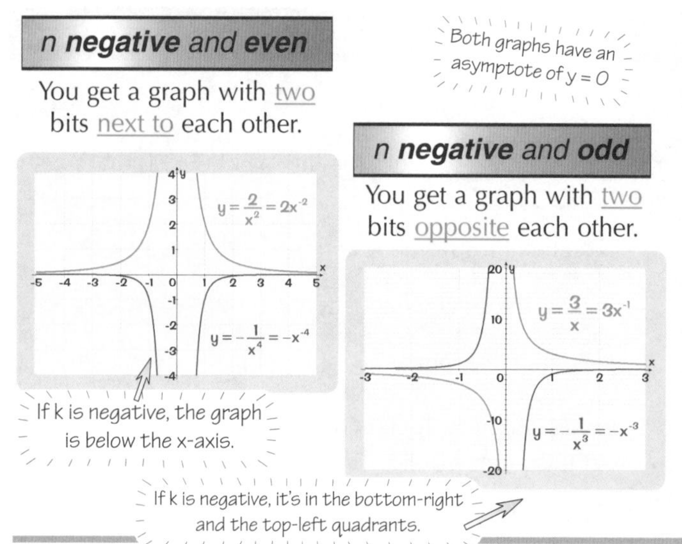

If you know the Factors of a cubic — the graph's easy to Sketch

A cubic function has an x^3 term in it, and all cubics have 'bottom-left to top-right' shape — or a 'top-left to bottom-right' shape if the coefficient of x^3 is negative.

If you know the factors of a cubic, the graph is easy to sketch — just find where the function is zero.

Example: Sketch the graphs of the following cubic functions.

(i) $f(x) = x(x-1)(2x+1)$ (ii) $g(x) = (1-x)(x^2-2x+2)$ (iii) $h(x) = (x-3)^2(x+1)$ (iv) $m(x) = (2-x)^3$

(i) The function's zero when $x = 0, 1$ or $-\frac{1}{2}$.

(ii) Differentiate — and you find the gradient's never zero. The coefficient of x^3 is negative, and the quadratic factor of g(x) has no roots — so g(x) is only zero once.

(iii) This has a 'double-root' at x = 3, so the graph just touches the x-axis there but doesn't go through.

(iv) A triple-root looks like this. This has a 'triple-root' at x = 2, and the coefficient of x^3 is negative.

Graphs, graphs, graphs — you can never have too many graphs...

It may seem like a lot to remember, but graphs can really help you get your head round a question — a quick sketch can throw a helluva lot of light on a problem that's got you completely stumped. So being able to draw these graphs won't just help with an actual graph-sketching question — it could help with loads of others too. Got to be worth learning.

Graph Transformations

Suppose you start with any old function f(x). Then you can transform (change) it in three ways — by translating it, stretching or reflecting it.

y = f(x)

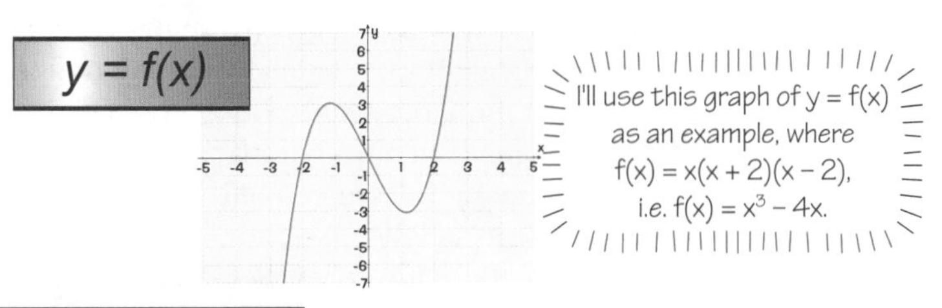

Translations are caused by Adding things

y = f(x)+a

Adding a number to the whole function shifts the graph up or down.

1) If $a > 0$, the graph goes upwards.
2) If $a < 0$, the graph goes downwards.

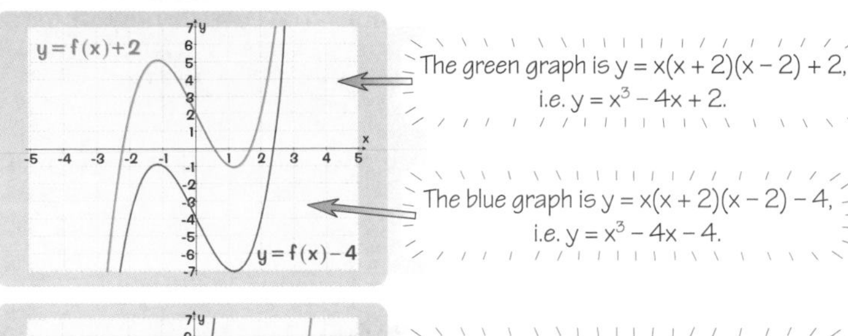

y = f(x+a)

Writing 'x + a' instead of 'x' means the graph moves sideways.

1) If $a > 0$, the graph goes to the left.
2) If $a < 0$, the graph goes to the right.

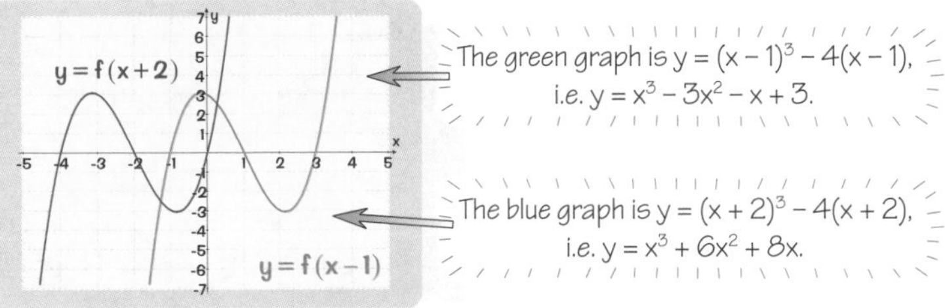

Stretches and Reflections are caused by Multiplying things

y = af(x)

Multiplying the whole function stretches, squeezes or reflects the graph vertically.

1) Negative values of 'a' reflect the basic shape in the x-axis.
2) If $a > 1$ or $a < -1$ (i.e. $|a| > 1$) the graph is stretched vertically.
3) If $-1 < a < 1$ (i.e. $|a| < 1$) the graph is squashed vertically.

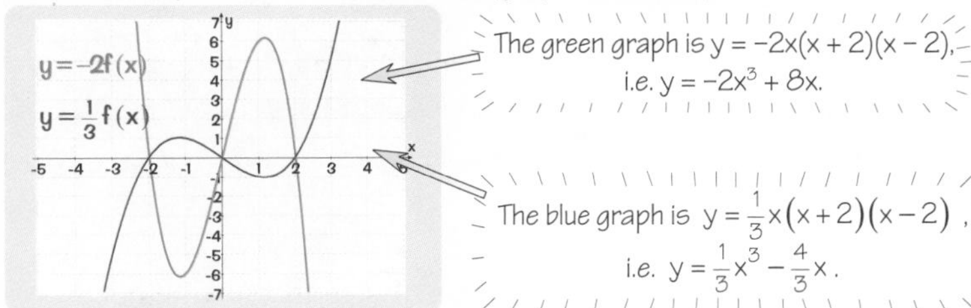

y = f(ax)

Writing 'ax' instead of 'x' stretches, squeezes or reflects the graph horizontally.

1) Negative values of 'a' reflect the basic shape in the y-axis.
2) If $a > 1$ or $a < -1$ (i.e. if $|a| > 1$) the graph is squashed horizontally.
3) If $-1 < a < 1$ (i.e. if $|a| < 1$) the graph is stretched horizontally.

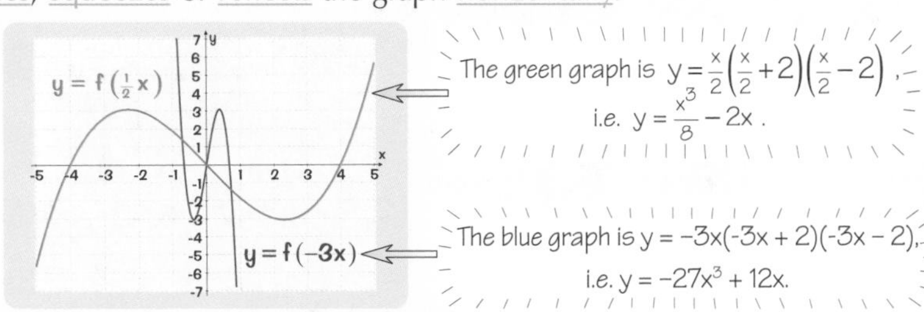

More than one transformation at a time: y = af(bx + c)+d

Example: $y = af(bx + c) + d$

$a = \frac{1}{3}$ ⇒ The graph is squashed vertically.

$b = \frac{1}{2}$ ⇒ The graph is stretched horizontally.

$c = \frac{1}{2}$ ⇒ The graph is moved horizontally (to the left).

$d = 1$ ⇒ The graph is shifted vertically (upwards).

This is a combination of all these transformations together.

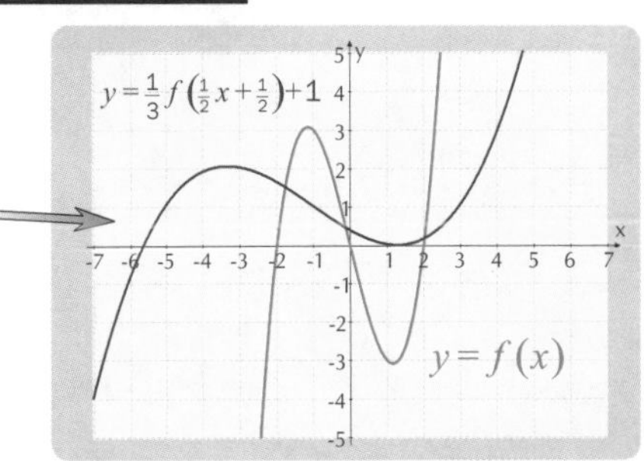

Section Four Revision Questions

There you go then... a section on various geometrical things. And in a way it was quite exciting, I'm sure you'll agree. You got the standard and (some might say) slightly dull straight lines and parabolas, but also the (ever so slightly) more exciting circle. And as you are probably aware, we mathematicians take our excitement from wherever we can get it. That's the good thing about AS maths really — it teaches you to really look hard for excitement at all opportunities, because you know that it's not going to come around too often. Anyway, that's quite enough of me. I'll leave you alone now to savour the lovely questions below to see how much knowledge you've absorbed as a result of working through the section. If you get them all correct, give yourself a pat on the back.
If not, read the section again until you know where you went wrong, and try the questions again.

1) Find the equations of the straight lines that pass through the points

   a) (2, –1) and (–4, –19), b) $(0, -\frac{1}{3})$ and $(5, \frac{2}{3})$.

   Write each of them in the forms

   i) $y - y_1 = m(x - x_1)$ ii) $y = mx + c$ iii) $ax + by + c = 0$, where a, b and c are integers.

2) a) The line l has equation $y = \frac{3}{2}x - \frac{2}{3}$. Find the equation of the lovely, cuddly line parallel to l, passing through the point with coordinates (4, 2). Name this line Lilly.

   b) The line m (whose name is actually Mike) passes through the point (6, 1) and is perpendicular to $2x - y - 7 = 0$. What is the equation of m?

3) The coordinates of points R and S are (1, 9) and (10, 3) respectively. Find the equation of the line perpendicular to RS, passing through the point (1, 9).

4) It's lovely, lovely curve sketching time — so draw rough sketches of the following curves:

   a) $y = -2x^4$, b) $y = \frac{7}{x^2}$, c) $y = -5x^3$, d) $y = -\frac{2}{x^5}$.

5) Admit it — you love curve-sketching. We all do — and like me, you probably can't get enough of it. So more power to your elbow, and sketch these cubic graphs:

   a) $y = (x - 4)^3$, b) $y = (3 - x)(x + 2)^2$, c) $y = (1 - x)(x^2 - 6x + 8)$, d) $y = (x - 1)(x - 2)(x - 3)$.

6) Right — now it's time to get serious. Put your thinking head on, and use the graph of f(x) to sketch what these graphs would look like after they've been 'transformed'.

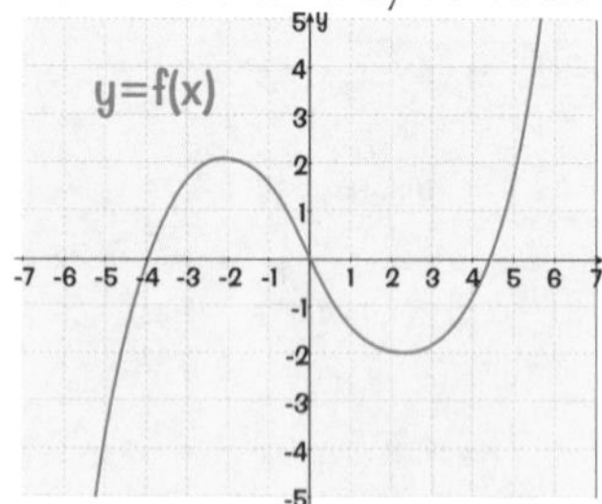

   a) $y = f(ax)$, where (i) $a > 1$, (ii) $0 < a < 1$,

   b) $y = af(x)$, where (i) $a > 1$, (ii) $0 < a < 1$,

   c) (i) $y = f(x + a)$, (ii) $y = f(x - a)$, where $a > 0$,

   d) (i) $y = f(x) + a$, (ii) $y = f(x) - a$, where $a > 0$.

Sequences

A sequence is a list of numbers that follow a certain pattern. Sequences can be finite or infinite (infinity — oooh), and they're usually generated in one of two ways. And guess what? You have to know everything about them.

A Sequence can be defined by its nth Term

You almost definitely covered this stuff at GCSE, so no excuses for mucking it up.
The point of all this is to show how you can work out any value (the nth term) from its position in the sequence (n).

Example: Find the nth term of the sequence 5, 8, 11, 14, 17, ...

1st	2nd	3rd	4th	5th
5	8	11	14	17

+3 +3 +3 +3

Each term is 3 more than the one before it. That means that you need to start by multiplying n by 3.
Take the first term (where $n = 1$). If you multiply n by 3, you still have to add 2 to get 5.
The same goes for $n = 2$. To get 8 you need to multiply n by 3, then add 2.
Every term in the sequence is worked out exactly the same way.

So n^{th} term is $3n + 2$

You can define a sequence by a Recurrence Relation too

Don't be put off by the fancy name — recurrence relations are pretty easy really.

The main thing to remember is:
a_k just means the kth term of the sequence

The next term in the sequence is a_{k+1}. You need to describe how to work out a_{k+1} if you're given a_k.

Example: Find the recurrence relation of the sequence 5, 8, 11, 14, 17, ...

From the example above, you know that each term equals the one before it, plus 3.

This is written like this: $\mathbf{a_{k+1} = a_k + 3}$

So, if $k = 5$, $a_k = a_5$ which stands for the 5th term, and $a_{k+1} = a_6$ which stands for the 6th term.

In everyday language, $a_{k+1} = a_k + 3$ means that the sixth term equals the fifth term plus 3.

BUT $a_{k+1} = a_k + 3$ on its own isn't enough to describe 5, 8, 11, 14, 17,...
For example, the sequence 87, 90, 93, 96, 99, ... also has each term being 3 more than the one before.

The recurrence relation needs to be more specific, so you've got to give one term in the sequence.
You almost always give the first value, a_1.

Putting all of this together gives 5, 8, 11, 14, 17,... as $a_{k+1} = a_k + 3$, $a_1 = 5$

Arithmetic Progressions

Right, you've got basic sequences tucked under your belt now — time to step it up a notch (sounds painful). When the terms of a sequence progress by adding a fixed amount each time, this is called an arithmetic progression.

It's all about Finding the nth Term

The first term of a sequence is given the symbol **a**. The amount you add each time is called the common difference, called **d**. The position of any term in the sequence is called **n**.

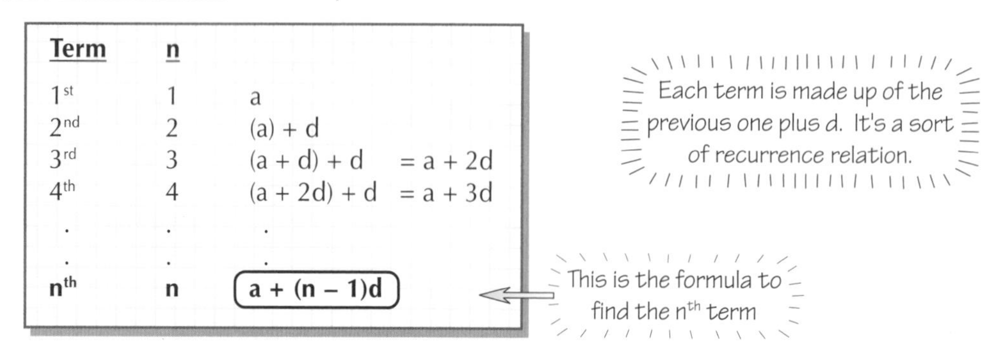

Term	n		
1st	1	a	
2nd	2	$(a) + d$	
3rd	3	$(a + d) + d$	$= a + 2d$
4th	4	$(a + 2d) + d$	$= a + 3d$
.	.	.	
.	.	.	
n^{th}	n	$a + (n - 1)d$	

Example: Find the 20th term of the arithmetic progression 2, 5, 8, 11,... and find the formula for the nth term.

Here $a = 2$ and $d = 3$

To get d, just find the difference between two terms next to each other — e.g. 11 – 8 = 3

So 20th term $= a + (20 - 1)d$
$= 2 + 19 \times 3$
$= 59$

The general term is the nth term, i.e. $a + (n - 1)d$
$= 2 + (n - 1)3$
$= 3n - 1$

A Series is when you Add the Terms to Find the Total

S_n is the total of the first n terms of the arithmetic progression:

$$S_n = a + (a + d) + (a + 2d) + (a + 3d) + \ldots + (a + (n - 1)d)$$

There's a really neat version of the same formula too:

$$S_n = n \times \frac{(a + l)}{2}$$

The l stands for the last value in the progression. You work it out as $l = a + (n - 1)d$

Nobody likes formulas, so think of it as the average of the first and last terms multiplied by the number of terms.

Example: Find the sum of the series with first term 3, last term 87 and common difference 4.

Here you know a, d and l, but you don't know n yet.

Use the information about the last value, l: $a + (n - 1)d = 87$
Then plug in the other values: $3 + 4(n - 1) = 87$

$4n - 4 = 84$

$4n = 88$

$n = 22$

So $S_{22} = 22 \times \frac{(3+87)}{2}$

$S_{22} = 990$

Arithmetic Series and Sigma Notation

They *Won't* always give you the *Last Term*...

...but don't panic — there's a formula to use when the last term is unknown. But you knew I'd say that, didn't you?

You know $l = a + (n - 1)d$ and $S_n = n\frac{(a+l)}{2}$.

Plug l into S_n and rearrange to get the formula in the box:

$$S_n = \frac{n}{2}[2a+(n-1)d]$$

Example: For the sequence -5, -2, 1, 4, 7, ... find the sum of the first 20 terms.

So $a = -5$ and $d = 3$.
The question says $n = 20$ too.

$$S_{20} = \frac{20}{2}[2\times -5+(20-1)\times 3]$$
$$= 10\,[-10+19\times 3]$$
$$S_{20} = 470$$

There's *Another* way of *Writing Series, too*

So far, the letter S has been used for the sum. The Greeks did a lot of work on this — their capital letter for S is sigma, or Σ. This is used today, together with the general term, to mean the sum of the series.

...and ending with n=15

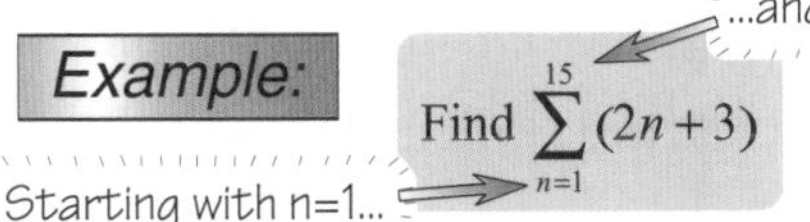

This means you have to find the sum of the first 15 terms of the series with n^{th} term $2n + 3$.

The first term ($n = 1$) is 5, the second term ($n = 2$) is 7, the third is 9, ... and the last term ($n = 15$) is 33. In other words, you need to find 5 + 7 + 9 + ... + 33. This gives $a = 5$, $d = 2$, $n = 15$ and $l = 33$.

You know all of a, d, n and l, so you can use either formula:

$$S_n = n\frac{(a+l)}{2}$$
$$S_{15} = 15\frac{(5+33)}{2}$$
$$S_{15} = 15 \times 19$$
$$S_{15} = 285$$

It makes no difference which method you use.

$$S_n = \frac{n}{2}[2a + (n - 1)d]$$
$$S_{15} = \frac{15}{2}[2 \times 5 + 14 \times 2]$$
$$S_{15} = \frac{15}{2}[10 + 28]$$
$$S_{15} = 285$$

Use *Arithmetic Progressions* to add up the *First n Whole Numbers*

The sum of the first n natural numbers looks like this:

$$S_n = 1 + 2 + 3 + \ldots + (n-2) + (n-1) + n$$

So $a = 1$, $l = n$ and also $n = n$.
Now just plug those values into the formula:

$$S_n = n \times \frac{(a+l)}{2} \Longrightarrow S_n = \frac{1}{2}n(n+1)$$

Natural numbers are just positive whole numbers.

Example:

Add up all the whole numbers from 1 to 100.

Sounds pretty hard, but all you have to do is stick it into the formula:

$S_{100} = ½ \times 100 \times 101$. So $S_{100} = 5050$

This sigma notation is all Greek to me... (Ho ho ho)

A sequence is just a list of numbers (with commas between them) — a series on the other hand is when you add all the terms together. It doesn't sound like a big difference, but mathematicians get all hot under the collar when you get the two mixed up. Remember that BlackADDer was a great TV series — not a TV sequence. (Sounds daft, but I bet you remember it now.)

Section Five Revision Questions

Well there's not a whole heap to learn in this section, but it's still important you work through these revision questions just to make sure you're as happy as a ferret in a trouser shop with Sequences and Series. Dead easy marks to pick up in the exam — so you'd feel pretty daft if you didn't learn this stuff. Come on then, get practising...

1) Find the nth term for the following sequences:

   a) 2, 6, 10, 14, ...

   b) 0.2, 0.7, 1.2, 1.7, ...

   c) 21, 18, 15, 12, ...

   d) 76, 70, 64, 58, ...

2) Find the recurrence relation of the sequence 32, 37, 42, 47, ...

3) Find the last term of the sequence that starts with 3, has a common difference of 0.5 and has 25 terms.

4) Find the common difference in a sequence that starts with -2, ends with 19 and has 29 terms

5) Find the sum of the series that starts with 7, ends with 35 and has 8 terms.

6) Find the sum of the series that begins with 5, 8, … and ends with 65.

7) A series has first term 7 and 5th term 23.

   Find: a) the common difference, b) the 15th term, and c) the sum of the first ten terms.

8) A series has seventh term 36 and tenth term 30. Find the sum of the first five terms and the nth term.

9) Find $\sum_{n=1}^{20}(3n-1)$

10) Find $\sum_{n=1}^{10}(48-5n)$

Differentiation

Brrrrrr... differentiation is a bad one — it really is. Not because it's that hard, but because it comes up all over the place in exams. So if you don't know it perfectly, you're asking for trouble.
Differentiation is just a way to work out gradients of graphs. You take a function, differentiate it, and you can quickly tell how steep a graph is. It's magic.

$$\frac{d}{dx}\left(x^n\right) = nx^{n-1}$$

$\frac{d}{dx}$ just means 'the derivative of the thing in the brackets'.

Use this formula to differentiate Powers of x

Equations are much easier to differentiate when they're written as powers of x — like writing $\sqrt{x}$ as $x^{\frac{1}{2}}$.
When you've done this, you can use the formula (the thing in the red box above) to differentiate the equation.

Use the differentiation formula...

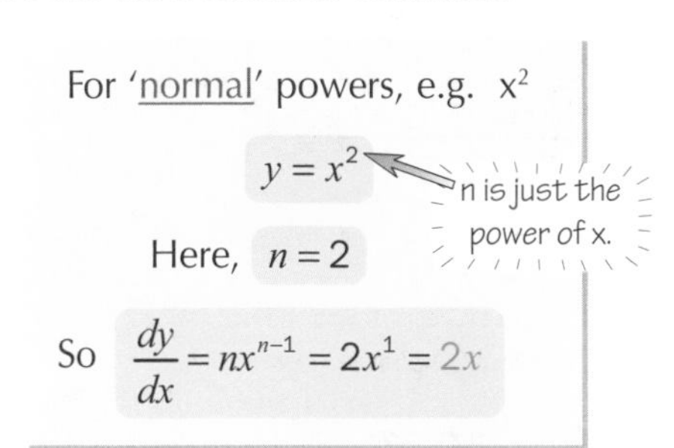

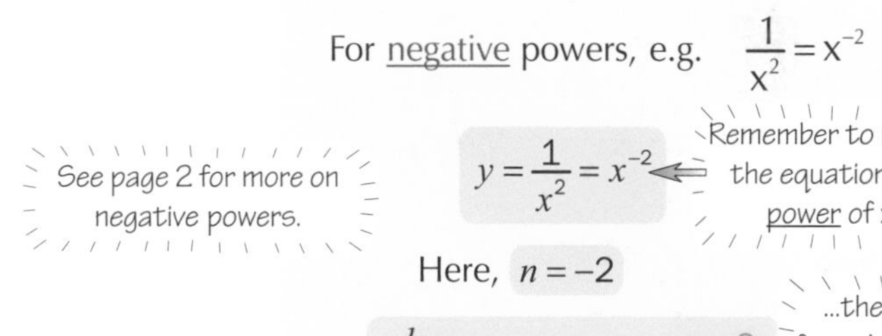

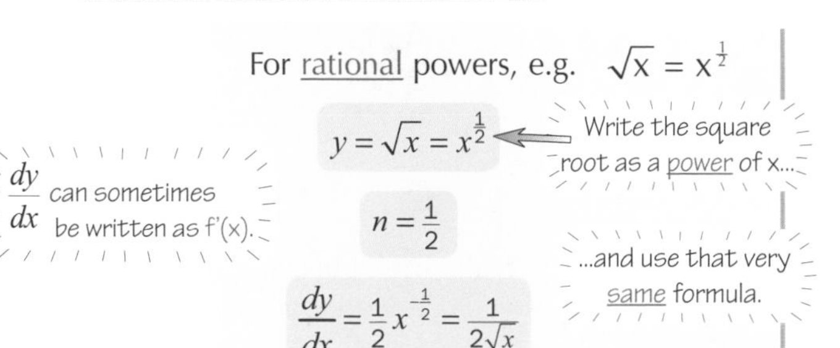

Power Laws:

Differentiation's much easier if you know the Power Laws really well. Like knowing that $x^1 = x$ and $\sqrt{x} = x^{\frac{1}{2}}$. See page 2 for more info.

Differentiate each term in an equation Separately

This formula is better than cake — even better than that really nice sticky black chocolate one from that place in town. Even if there are loads of terms in the equation, it doesn't matter. Differentiate each bit separately and you'll be fine.

Here are a couple of examples...

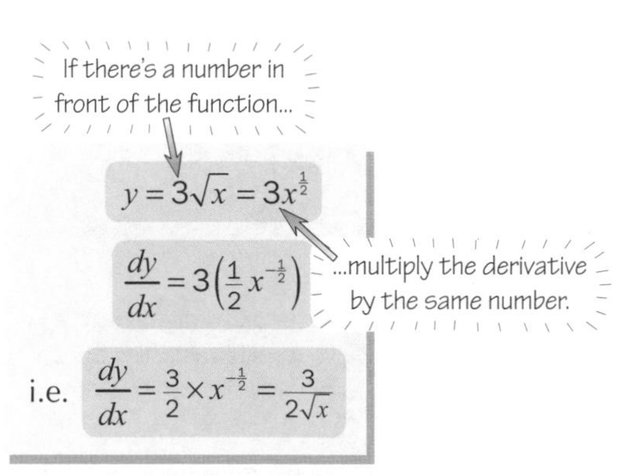

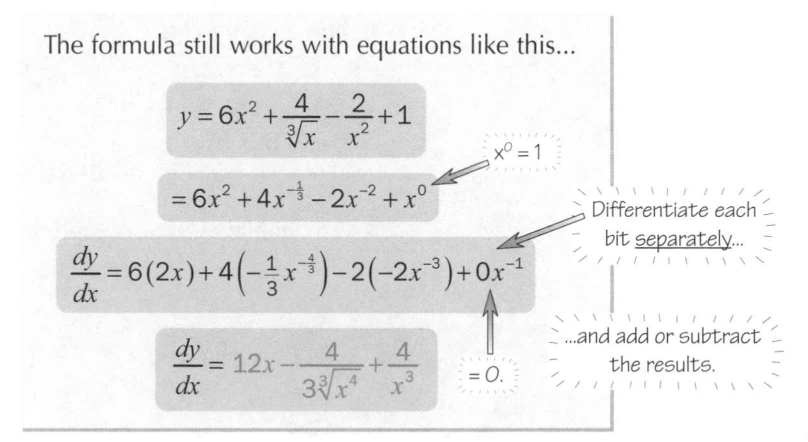

Dario O'Gradient — differentiating Crewe from the rest...

If you're going to bother doing maths, you've got to be able to differentiate things. Simple as that. But luckily, once you can do the simple stuff, you should be all right. Big long equations are just made up of loads of simple little terms, so they're not really that much harder. Learn the formula, and make sure you can use it by practising all day and all night forever.

Differentiation

Differentiation's what you do if you need to find a gradient. Excited yet?

Differentiate to find Gradients...

EXAMPLE: Find the gradient of the graph $y = x^2$ at $x = 1$ and $x = -2$...

You need the gradient of the graph of...

$$y = x^2$$

So differentiate this function to get...

$$\frac{dy}{dx} = 2x$$

Now when $x = 1$, $\frac{dy}{dx} = 2$

And so the gradient of the graph at x = 1 is 2.

And when $x = -2$, $\frac{dy}{dx} = -4$.

So the gradient of the graph at x = –2 is –4.

Use differentiation to find the gradient of a curve — which is the same as the gradient of the tangent at that point.

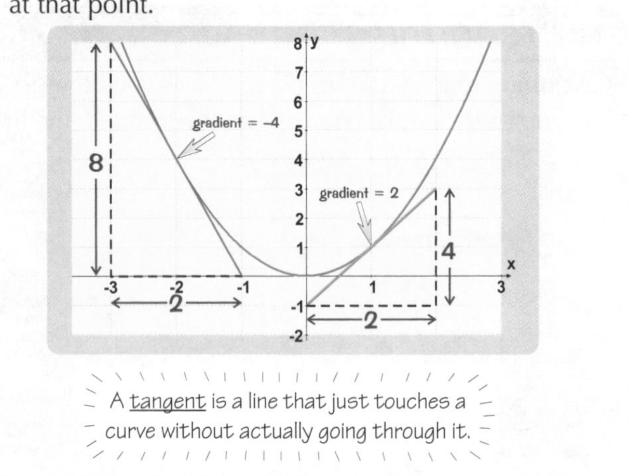

A tangent is a line that just touches a curve without actually going through it.

...which tell you Rates of Change...

So, you've differentiated an equation and found the gradient at a point — which is really useful because this tells you the rate of change of the curve at that point (e.g. from distance vs. time graphs you can work out speed).

EXAMPLE: A sports car pulls off from a junction and drives away, travelling a distance (*d* metres), in time (*t* seconds). For the first 10 seconds, its path can be described by the equation $d = 2t^2$

Find: a) the speed of the car after 8 seconds and b) the car's acceleration during this period.

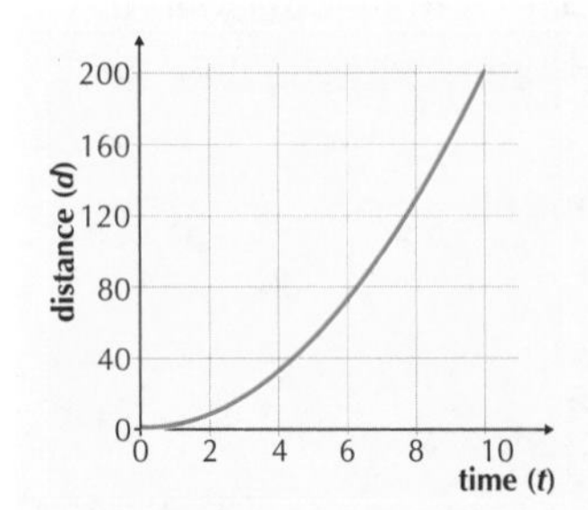

a) You can work out the speed by measuring the gradient of the curve $d = 2t^2$.

Differentiate to give: $\frac{dd}{dt} = 4t$ When t = 8, $\frac{dd}{dt} = 32$

So, the car is travelling at 32 m/s after 8 seconds.

b) Acceleration is the rate that speed (s) changes (i.e. it is the gradient of $s = 4t$).

so differentiate again to find the acceleration: $\frac{d^2d}{dt^2} = 4$

This is called a second order derivative — because you've differentiated twice

Help me Differentiation — You're my only hope...

There's not much hard maths on this page — but there are a couple of very important ideas that you need to get your head round pretty darn soon. Understanding that differentiating gives the gradient of the graph is more important than washing regularly — AND THAT'S IMPORTANT. The other thing on the page you need to know is that gradient tells you the rate of change of a function — which is also vital when working out what a question is after.

Finding Tangents and Normals

What's a tangent? Beats me. Oh no, I remember, it's one of those thingies on a curve. Ah, yes... I remember now...

Tangents Just touch a curve

To find the equation of a tangent or a normal to a curve, you first need to know its gradient — so differentiate. Then complete the line's equation using the coordinates of one point on the line.

Find the tangent to the curve y = (4 – x)(x + 2) at the point (2, 8).

Tangents and Normals...

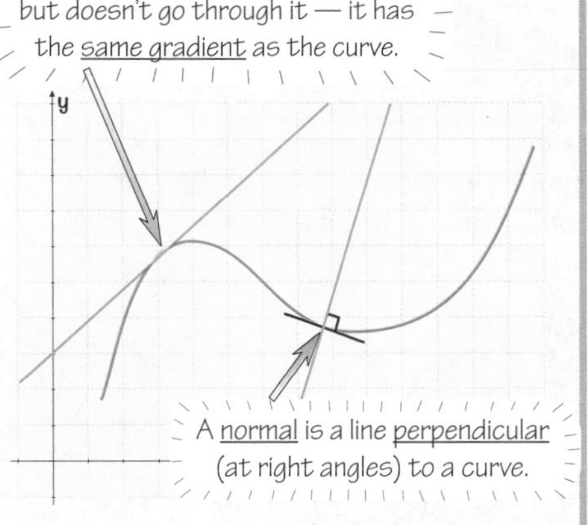

To find the curve's (and the tangent's) gradient, first write the equation in a form you can differentiate...

$$y = 8 + 2x - x^2$$

...and then differentiate it.

$$\frac{dy}{dx} = 2 - 2x$$

The gradient of the tangent will be the gradient of the curve at x = 2.

At $x = 2$, $\frac{dy}{dx} = -2$,

So the tangent has equation,

$$y - y_1 = -2(x - x_1)$$

in $y - y_1 = m(x - x_1)$ form. See page 26.

And since it passes through the point (2,8), this becomes

$y - 8 = -2(x - 2)$, or $y = -2x + 12$.

You can also write it in $y = mx + c$ form.

Normals are at Right Angles to a curve

EXAMPLE: Find the normal to the curve $y = \frac{(x+2)(x+4)}{6\sqrt{x}}$ at the point (4, 4).

Write the equation of the curve in a form you can differentiate.

$$y = \frac{x^2 + 6x + 8}{6x^{\frac{1}{2}}} = \frac{1}{6}x^{\frac{3}{2}} + x^{\frac{1}{2}} + \frac{4}{3}x^{-\frac{1}{2}}$$

Dividing everything on the top line by everything on the bottom line.

Differentiate it...

$$\frac{dy}{dx} = \frac{1}{6}\left(\frac{3}{2}x^{\frac{1}{2}}\right) + \frac{1}{2}x^{-\frac{1}{2}} + \frac{4}{3}\left(-\frac{1}{2}x^{-\frac{3}{2}}\right)$$
$$= \frac{1}{4}\sqrt{x} + \frac{1}{2\sqrt{x}} - \frac{2}{3\sqrt{x^3}}$$

Find the gradient at the point you're interested in. At x = 4,

$$\frac{dy}{dx} = \frac{1}{4} \times 2 + \frac{1}{2 \times 2} - \frac{2}{3 \times 8} = \frac{2}{3}$$

So the gradient of the normal is $-\frac{3}{2}$.

Because the gradient of the normal multiplied by the gradient of the curve must be -1.

And the equation of the normal is $y - y_1 = -\frac{3}{2}(x - x_1)$.

Finally, since the normal goes through the point (4, 4), the equation of the normal must be $y - 4 = -\frac{3}{2}(x - 4)$, or after rearranging, $y = -\frac{3}{2}x + 10$.

Finding tangents and normals

1) **Differentiate the function.**
2) **Find the gradient, m, of the tangent or normal. This is,**
   for a tangent: the gradient of the curve
   for a normal: $\frac{-1}{\text{gradient of the curve}}$
3) **Write the equation of the tangent or normal in the form $y - y_1 = m(x - x_1)$, or $y = mx + c$.**
4) **Complete the equation of the line using the coordinates of a point on the line.**

Repeat after me... "I adore tangents and normals..."

Examiners can't stop themselves saying the words 'Find the tangent...' and 'Find the normal...'. They love the words. These phrases are music to their ears. They can't get enough of them. I just thought it was my duty to tell you that. And so now you know, you'll definitely be wanting to learn how to do the stuff on this page. Of course you will.

Section Six Revision Questions

That's what differentiation is all about. And frankly, it probably isn't the worst topic you'll meet in AS maths. Yes, there are fiddly things to remember — but overall, it's not as bad as all that. And just think of all the lovely marks you'll get if you can answer questions like these in the exam...

1) An easy one to start with. Write down the formula for differentiating any power of x.

2) Differentiate these functions with respect to x:

a) $y = x^2 + 2$,

b) $y = x^4 + \sqrt{x}$,

c) $y = \frac{7}{x^2} - \frac{3}{\sqrt{x}} + 12x^3$

3) What's the connection between the gradient of a curve at a point and the gradient of the tangent to the curve at the same point? (That sounds like a joke in need of a punchline — but sadly, this is no joke.)

4) Find the gradients of these graphs at x = 2

a) 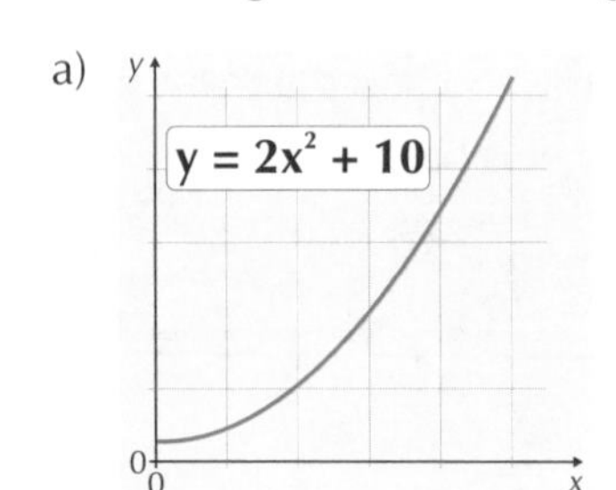

b) 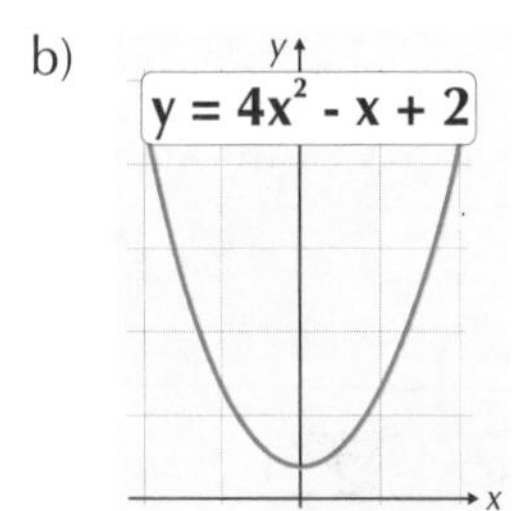

c) 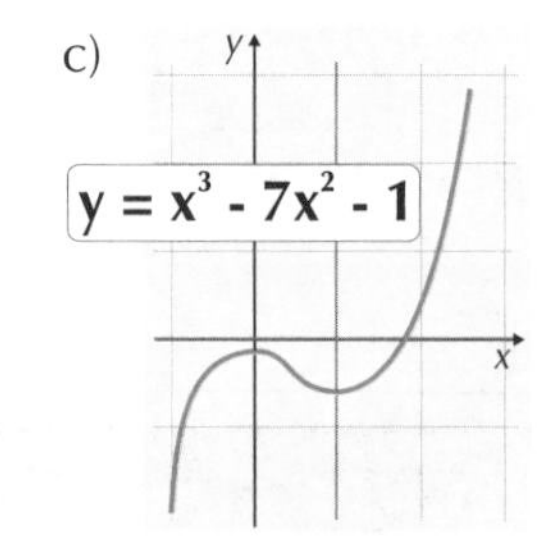

5) 1 litre of water is poured into a bowl.
The volume (v) of water in the bowl (in ml) is defined by the function: $v = 17t^2 - 10t$
Find the rate at which water is poured into the bowl when $t = 4$ seconds.

6) Yawn, yawn. Find the equations of the tangent and the normal to the curve $y = \sqrt{x^3} - 3x - 10$ at x = 16.

7) Show that the lines $y = \frac{x^3}{3} - 2x^2 - 4x + \frac{86}{3}$ and $y = \sqrt{x}$ both go through the point (4,2), and are perpendicular at that point. Good question, that — nice and exciting, just the way you like 'em.

Integration

Integration is the 'opposite' of differentiation — and so if you can differentiate, you can be pretty confident you'll be able to integrate too. There's just one extra thing you have to remember — the constant of integration...

You need the constant because there's More Than One right answer

When you integrate something, you're trying to find a function that returns to what you started with when you differentiate it. And when you add the constant of integration, you're just allowing for the fact that there's more than one possible function that does this...

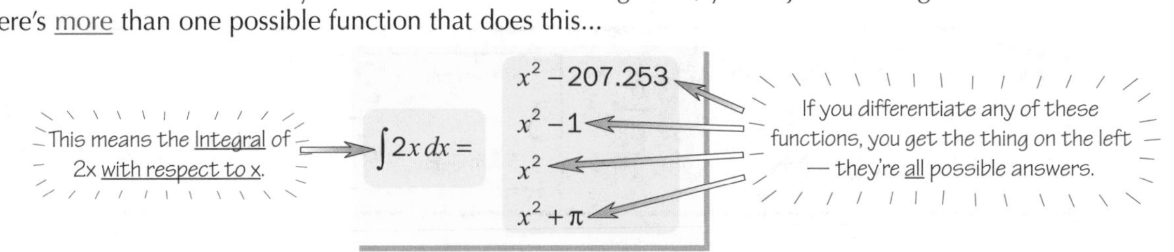

So the answer to this integral is actually...

$$\int 2x\,dx = x^2 + C$$

The 'C' just means 'any number'. This is the constant of integration.

You only need to add a constant of integration to indefinite integrals like these ones. Definite integrals are just integrals with limits (or little numbers) next to the integral sign, but you won't need to know about these in this module.

Up the power by One — then Divide by it

The formula below tells you how to integrate any power of x (except x^{-1}).

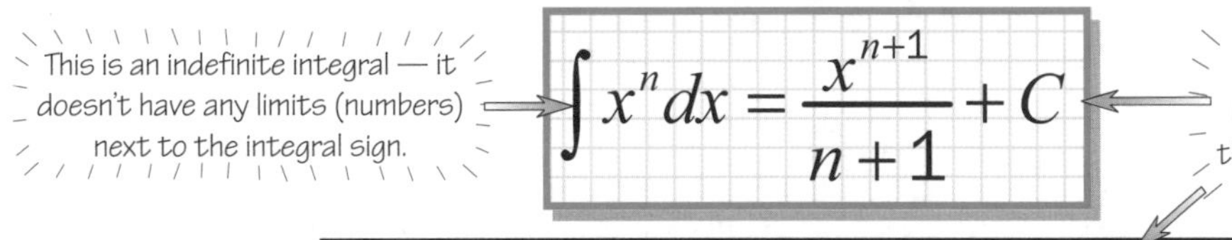

$$\int x^n\,dx = \frac{x^{n+1}}{n+1} + C$$

You can't do this to $\frac{1}{x} = x^{-1}$. When you increase the power by 1 (to get zero) and then divide by zero — you get big problems.

In a nutshell, this says:

To integrate a power of x: (i) Increase the power by one — then divide by it.
and (ii) Stick a constant on the end.

EXAMPLES: Use the integration formula...

① For 'normal' powers,

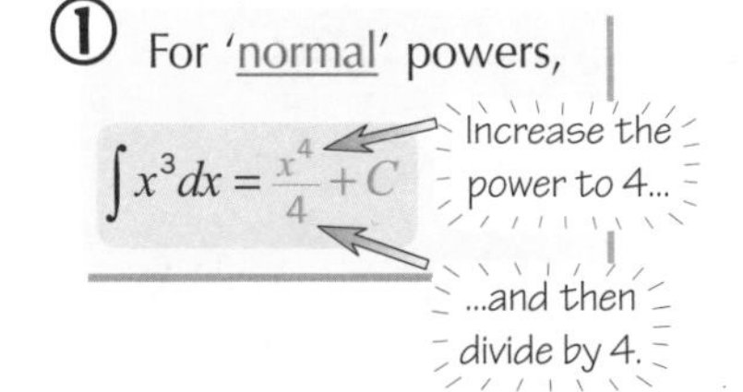

② For negative powers,

$$\int \frac{1}{x^3}dx = \int x^{-3}dx$$
$$= \frac{x^{-2}}{-2} + C$$
$$= -\frac{1}{2x^2} + C$$

Increase the power by 1 to −2...

...and then divide by −2.

③ For fractional powers,

$$\int \sqrt[3]{x^4}\,dx = \int x^{\frac{4}{3}}dx$$
$$= \frac{x^{\frac{7}{3}}}{(7/3)} + C$$
$$= \frac{3\sqrt[3]{x^7}}{7} + C$$

Add 1 to the power...

...then divide by this new power.

④ And for complicated looking stuff...

$$\int\left(3x^2 - \frac{2}{\sqrt{x}} + \frac{7}{x^2}\right)dx = \int\left(3x^2 - 2x^{-\frac{1}{2}} + 7x^{-2}\right)dx$$
$$= \frac{3x^3}{3} - \frac{2x^{\frac{1}{2}}}{(1/2)} + \frac{7x^{-1}}{-1} + C$$
$$= x^3 - 4\sqrt{x} - \frac{7}{x} + C$$

Do each of these bits separately.

CHECK YOUR ANSWERS:
You can check you've integrated properly by differentiating the answer — you should end up with the thing you started with.

Indefinite integrals — joy without limits...

This integration lark isn't so bad then — there's only a couple of things to remember and then you can do it no problem. But that constant of integration catches loads of people out — it's so easy to forget — and you'll definitely lose marks if you do forget it. You have been warned. Other than that, there's not much to it. Hurray.

Integration

By now, you're probably aware that maths isn't something you do unless you're a bit of a thrill-seeker. You know, sometimes they even ask you to find a curve with a certain derivative that goes through a certain point.

You sometimes need to find the Value of the Constant of Integration

When they tell you something else about the curve in addition to its derivative, you can work out the value of that constant of integration. Usually the something is the coordinates of one of the points the curve goes through.

Really Important Bit...

When you differentiate y, you get $\frac{dy}{dx}$.

And when you integrate $\frac{dy}{dx}$, you get y*.

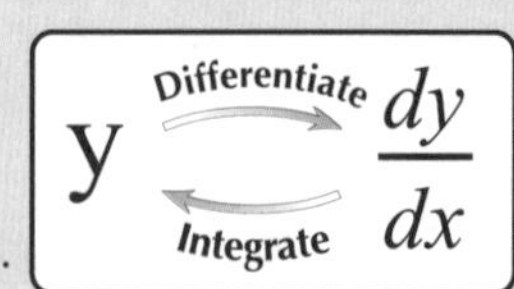

*If you ignore the constant of integration.

EXAMPLE: Find the equation of the curve through the point (2, 8) with $\frac{dy}{dx} = 6x(x-1)$.

> **Remember:**
> Even if you don't have any extra information about the curve — you still have to add a constant when you work out an integral without limits.

You know the derivative and need to find the function — so integrate.

$$\frac{dy}{dx} = 6x(x-1) = 6x^2 - 6x$$

So integrating both sides gives...

$$y = \int (6x^2 - 6x)\,dx$$
$$\Rightarrow y = \frac{6x^3}{3} - \frac{6x^2}{2} + C$$
$$\Rightarrow y = 2x^3 - 3x^2 + C$$

Don't forget the constant of integration.

Check this is correct by differentiating it and making sure you get what you started with.

$$y = 2x^3 - 3x^2 + C = 2x^3 - 3x^2 + Cx^0$$
$$\Rightarrow \frac{dy}{dx} = 2(3x^2) - 3(2x^1) + C(0x^{-1})$$
$$\Rightarrow \frac{dy}{dx} = 6x^2 - 6x$$

So this function's got the correct derivative — but you haven't finished yet.

You now need to find C — and you do this by using the fact that it goes through the point (2, 8).

$$y = 2x^3 - 3x^2 + C$$

Putting x = 2 and y = 8 in the above equation gives...

$$8 = (2\times2^3) - (3\times2^2) + C$$
$$\Rightarrow 8 = 16 - 12 + C$$
$$\Rightarrow C = 4$$

So the answer you need is this one:

$$y = 2x^3 - 3x^2 + 4$$

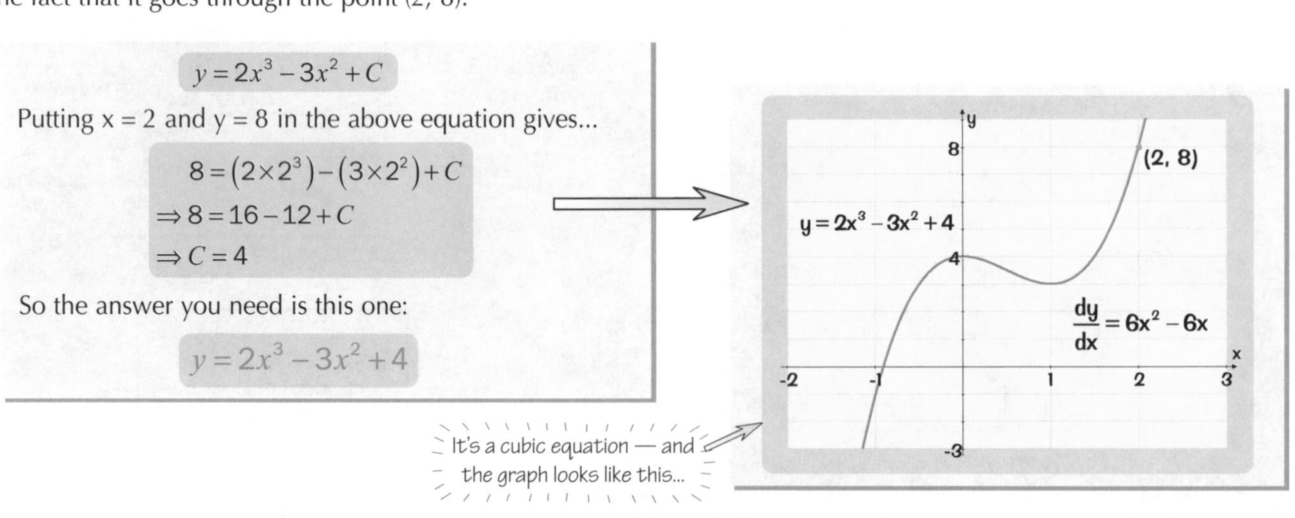

It's a cubic equation — and the graph looks like this...

Maths and alcohol don't mix — so never drink and derive...

That's another page under your belt and — go on, admit it — there was nothing too horrendous on it. If you can do the stuff from the previous page and then substitute some numbers into an equation, you can do everything from this page too. So if you think this is boring, you'd be right. But if you think it's much harder than the stuff before, you'd be wrong.

Section Seven Revision Questions

Integration isn't a whole heap different from differentiation really. Well, that's not true — integration is pretty much the *opposite* of differentiation, so in that sense it's completely different. But if you can differentiate, then I'd feel pretty confident that you could integrate as well. Which brings us (kind of) neatly on to these questions. If you've read the section and feel ready to test your integration knowledge, then have a go at the questions below. You need to be aiming to get all of them right. But if you do make a mistake, then it's not the end of the world — just re-read the relevant part of the section and then have another go. And keep doing this until you don't make any mistakes. Then you can feel ready to take on any integration exam questions that Core 1 might throw at you.

1) Write down the steps involved in integrating a power of x.

2) What's an indefinite integral? Why do you have to add a constant of integration when you find an indefinite integral?

3) How can you check whether you've integrated something properly? (Without asking someone else.)

4) Integrate these: a) $\int 10x^4 dx$, b) $\int (3x + 5x^2) dx$, c) $\int \left(x^2(3x+2)\right) dx$

5) Work out the equation of the curve that goes through the point (1, 0) and has derivative $\frac{dy}{dx} = 6x - 7$.

6) Find the equation of the curve that has derivative $\frac{dy}{dx} = 3x^3 + 2$ and goes through the point (1, 0). How would you change the equation if the curve had to go through the point (1, 2) instead? (Don't start the whole question again.)

General Certificate of Education
Advanced Subsidiary (AS) and Advanced Level

Core 1 Mathematics — Practice Exam One

Give non-exact numerical answers correct to 3 significant figures, unless a different degree of accuracy is specified in the question or is clearly appropriate.

1 (i) Write down the exact value of $36^{-½}$ [2]

(ii) Simplify $\dfrac{a^6 \times a^3}{\sqrt{a^4}} \div a^{\frac{1}{2}}$ [2]

(iii) Express $\left(5\sqrt{5}+2\sqrt{3}\right)^2$ in the form $a+b\sqrt{c}$, where a, b and c are integers to be found. [4]

(iv) Rationalise the denominator of $\dfrac{10}{\sqrt{5}+1}$. [2]

2 (i) Either algebraically, or by sketching the graphs, solve the inequality $4x+7>7x+4$ [2]

(ii) Find the values of k, such that $(x-5)(x-3)>k$ for all possible values of x. [3]

(iii) Find the range of x that satisfies the inequality $(x+3)(x-2)<2$. [3]

3 (i) Find the coordinates of the point A, when A lies at the intersection of the lines l_1 and l_2, and when the equations of l_1 and l_2 respectively are $x-y+1=0$ and $2x+y-8=0$. [3]

(ii) The points B and C have coordinates (6, –4) and $\left(-\frac{4}{3},-\frac{1}{3}\right)$ respectively, and D is the midpoint of AC.

Find the equation of the line BD in the form ax + by + c = 0, where a, b and c are integers. [6]

(iii) Show that the triangle ABD is a right-angled triangle. [3]

4 Solve the following simultaneous equations:

$$y+x=7 \quad \text{and} \quad y=x^2+3x-5$$ [4]

5 (i) Express x^2-6x+5 in the form $(x+a)^2+b$. [2]

(ii) Factorise the expression x^2-6x+5. [2]

(iii) Hence sketch the graph of $y=x^2-6x+5$, clearly indicating the coordinates of the vertex and the points where it cuts the axes. [3]

6 A curve that passes through the point (2, 0) has derivative $dy/dx = 3x^2 + 6x - 4$.

(i) Show that the equation of the curve is $y = x^3 + 3x^2 - 4x - 12$. [3]

(ii) $(x + 3)$ is a factor of y. Express y as a product of 3 linear factors. [2]

7 The line AB is part of the line with equation $y + 2x - 5 = 0$.

A is the point with coordinates (1, 3) and B is the point with coordinates $(4, k)$.

(i) Find the value of k. [1]

(ii) What is the equation of the line perpendicular to AB, that passes through A? [3]

8 (i) An arithmetic series has first term a and common difference d.

(a) Write down expressions for u_n, u_{n+1} and u_{n+2}, the nth, $(n+1)^{th}$ and $(n+2)^{th}$ terms in the series respectively. [2]

(b) By making the substitution $x = a + nd$, write these expressions in terms of x and d only. [4]

(c) If the sum of u_n, u_{n+1} and u_{n+2} is 36, and their product is 960, find the positive values of x and d. [4]

(d) If u_n, u_{n+1} and u_{n+2} above are the first three terms of the series, write down an expression for u_n in terms of n. [3]

(ii) Find S_{10}, the sum of the first ten terms of the series. [3]

(iii) By considering the formula for S_n and the formula for the sum $\sigma_n = 1 + 2 + 3 + ... + n$, find an expression for the difference $S_n - \sigma_n$, giving your answer in as simple a form as possible. [4]

9 Find dy/dx for each of the following:

(i) $y = x^2$ [1]

(ii) $y = 3x^4 - 2x$ [2]

(iii) $y = (x^2 + 4)(x - 2)$ [2]

Paper 1 Q1 — Powers and Surds

1	**(i)**	Write down the exact value of $36^{-\frac{1}{2}}$	[2]
	(ii)	Simplify $\dfrac{a^6 \times a^3}{\sqrt{a^4}} \div a^{\frac{1}{2}}$	[2]
	(iii)	Express $(5\sqrt{5}+2\sqrt{3})^2$ in the form $a+b\sqrt{c}$, where a, b and c are integers to be found.	[4]
	(iv)	Rationalise the denominator of $\dfrac{10}{\sqrt{5}+1}$.	[2]

(i) *Power Law questions can usually be answered by Rearranging*

'Write down the exact value of $36^{-\frac{1}{2}}$.'

With Power Law questions, you usually just have to remember a couple of basic formulas, then do a couple of sums pretty darn carefully. If you've forgotten any of the Laws, they're all on page 2. So, on with the question...

First get rid of the minus in the exponent / power...

$x^{-n} = \dfrac{1}{x^n}$ ⟹ $36^{-\frac{1}{2}} = \dfrac{1}{36^{\frac{1}{2}}}$

Then deal with the $\frac{1}{2}$.

$\dfrac{1}{36^{\frac{1}{2}}} = \dfrac{1}{\sqrt{36}}$ ⟸ $x^{\frac{1}{n}} = \sqrt[n]{x}$

And finally...

Do a simple square root. ⟹ $\dfrac{1}{\sqrt{36}} = \dfrac{1}{6}$

(ii) *Simplifying just means Rearranging as well*

'Simplify $\dfrac{a^6 \times a^3}{\sqrt{a^4}} \div a^{\frac{1}{2}}$'

When you're simplifying powers, it's a good idea to get them all looking the same.
In this example, get the individual bits in the form a^n.

See page 2 for more info on the Power Laws.

First simplify the tricky bit on the bottom of the fraction...

$\sqrt[m]{a^n} = a^{\frac{n}{m}}$ ⟹ $\dfrac{a^6 \times a^3}{\sqrt{a^4}} \div a^{\frac{1}{2}} = \dfrac{a^6 \times a^3}{a^2} \div a^{\frac{1}{2}}$

Then rewrite this so that you're only multiplying things.

$a^6 \times a^3 \times a^{-2} \times a^{-\frac{1}{2}}$ ⟸ $\dfrac{1}{a^n} = a^{-n}$ — Dividing by a^n is the same as multiplying by a^{-n}.

And then just add all the powers together, to get

$a^6 \times a^3 \times a^{-2} \times a^{-\frac{1}{2}} = a^{6+3-2-\frac{1}{2}}$

$= a^{\frac{13}{2}}$ Hurray...

This may sound stupid, but questions on Power Laws aren't too bad, as long as you obey the Power Laws. It really is that simple.

CHECK YOUR ANSWER:

$\dfrac{a^6 \times a^3}{\sqrt{a^4}} \div a^{\frac{1}{2}} = a^{\frac{13}{2}}$ — Check your answer by substituting a value for a.

$\dfrac{2^6 \times 2^3}{\sqrt{2^4}} \div 2^{\frac{1}{2}} = 2^{\frac{13}{2}}$ — Work each power of two out separately.

$\dfrac{64 \times 8}{4} \div 1.41421 = 90.509$

$90.509 = 90.509$ — Keep lots of decimal places.

If both sides are equal, you've got the right answer.

Paper 1 Q1 — Powers and Surds

(iii) Multiply out the brackets — then use the rules for Surds

'Express $(5\sqrt{5}+2\sqrt{3})^2$ in the form $a+b\sqrt{c}$, where a, b and c are integers to be found.'

Yet again, you've got to simplify and rearrange the equation.

First of all (after the initial shock of "Arrrgghh — surds") you should get rid of the squared sign around the brackets.

Multiply out the brackets first:

$$(5\sqrt{5}+2\sqrt{3})^2=(5\sqrt{5}+2\sqrt{3})\times(5\sqrt{5}+2\sqrt{3})$$

$$=(5\sqrt{5})^2+2(5\sqrt{5}\times2\sqrt{3})+(2\sqrt{3})^2$$

Multiply this out like a normal quadratic.

This next bit's a tad confusing, I reckon. You've got three terms to deal with, and they're all a little bit nasty.

The first term is:

$$(5\sqrt{5})^2=5\sqrt{5}\times5\sqrt{5}=5\times5\times\sqrt{5}\times\sqrt{5}=5\times5\times5=125$$

The second term is:

$$2(5\sqrt{5}\times2\sqrt{3})=2\times5\times2\times\sqrt{5}\times\sqrt{3}=20\times\sqrt{15}=20\sqrt{15}$$

And the third term is:

$$(2\sqrt{3})^2=2\sqrt{3}\times2\sqrt{3}=2\times2\times\sqrt{3}\times\sqrt{3}=2\times2\times3=12$$

This is a times sign, not an add sign — so don't 'work out' the brackets.

Remember — the order doesn't matter when you multiply.

$\sqrt{5}\sqrt{3}=\sqrt{5\times3}=\sqrt{15}$. Whatever you do, don't add the numbers in the square root signs.

$\sqrt{3}\times\sqrt{3}=3$

Now you've done that, though, you're almost home and dry. All that's left to do is add the three terms together.

So the whole thing's equal to...

$$125+20\sqrt{15}+12=137+20\sqrt{15}$$

And since 137, 20 and 15 are all integers, this is the final answer to the question.

(iv) Rationalise by multiplying top and bottom lines by the Same Thing

'Rationalise the denominator of $\frac{10}{\sqrt{5}+1}$.'

Rationalising a denominator means getting rid of surds on the bottom line of a fraction. Sounds hard — but it's easy.

The first thing to do is rewrite the bottom line so that it's in the form $a+\sqrt{b}$ — with the surd after the + or – sign.

If the bottom line doesn't already have the surd after the + or –...

$$\frac{10}{\sqrt{5}+1}=\frac{10}{1+\sqrt{5}}$$

...just rewrite the bottom line.

Then if you've got $a+\sqrt{b}$ on the bottom line — you have to multiply top and bottom lines by $a-\sqrt{b}$, and vice versa.

Just change the sign before the surd — and multiply top and bottom lines by it.

$$\frac{10}{1+\sqrt{5}}=\frac{10}{1+\sqrt{5}}\times\frac{1-\sqrt{5}}{1-\sqrt{5}}$$

Since the bottom line is $1+\sqrt{5}$, you have to multiply both the top and bottom lines by $1-\sqrt{5}$.

$$=\frac{10(1-\sqrt{5})}{(1+\sqrt{5})(1-\sqrt{5})}$$

Use the difference of two squares: $(x+y)(x-y)=x^2-y^2$.

$$=\frac{10(1-\sqrt{5})}{1-5}$$

$$=\frac{10(1-\sqrt{5})}{-4}=\frac{-5(1-\sqrt{5})}{2}$$

Keep your Magic Power Laws close to you at all times...

I don't want to go on and on and start ranting but... hang on a bit, I do want to go on and on, and I do want to rant. The thing is that questions on the Power Laws always come up in the exams and are always worth a good few marks. And those few marks could make the difference between one grade and the next. Think about it — half an hour spent learning this page really well could move you up a grade. You know it makes sense...

Paper 1 Q2 — Inequalities...

2 **(i)** Either algebraically, or by sketching the graphs, solve the inequality

$$4x+7>7x+4$$ [2]

(ii) Find the values of k, such that

$$(x-5)(x-3)>k \text{ for all possible values of } x.$$ [3]

(iii) Find the range of x that satisfies the inequality

$$(x+3)(x-2)<2.$$ [3]

(i) A straightforward Linear Inequality

'Either algebraically, or by sketching the graphs...'

That opening makes it sound really tricky, but don't be fooled. Sketching graphs sounds much easier, but it's such a simple inequality that it's much quicker to just work it out:

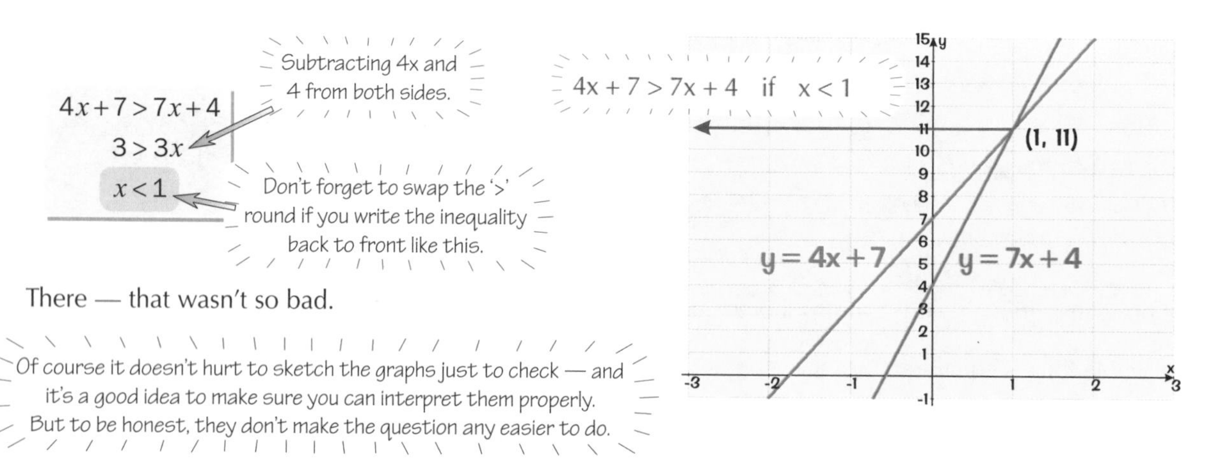

(ii) Easy — if you spot the Symmetry

'Find the values of k, such that

$(x-5)(x-3)>k$ for all possible values of x.'

Another way you could to this part would be to multiply out the brackets and complete the square.

Basically, you've got to find the minimum value of $(x-5)(x-3)$, and make sure k is less than that.

As always, if you're a bit unsure where to start, think what the function looks like — and SKETCH THE GRAPH.

Here's the cunning bit: The thing to realise is that the graph's symmetrical — so the minimum will be halfway between $x = 3$ and $x = 5$ — i.e. $x = 4$. So just plug that into the equation to find the lowest point the graph reaches:

Putting $x = 4$ in $(x-5)(x-3)$ gives

$$(4-5)(4-3) = -1\times1$$
$$= -1$$

So if $k<-1$, the graph will never be as low as k.

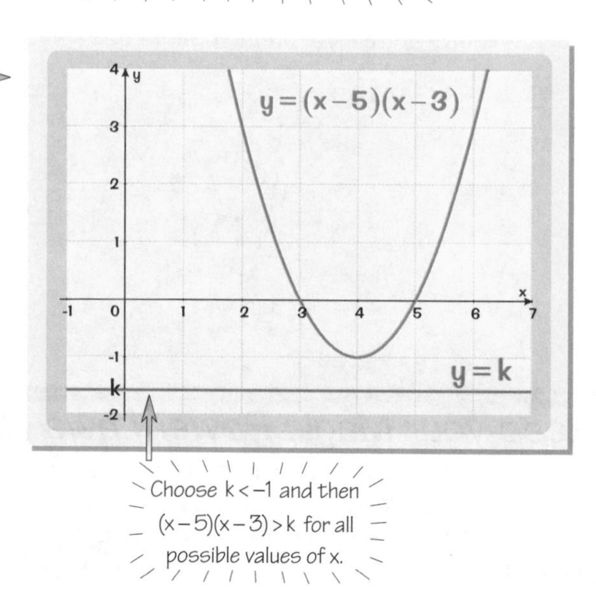

Paper 1 Q2 — ...and Quadratics

(iii) Draw the Graph to see when it's Negative

'Find the range of x that satisfies the inequality

$(x + 3)(x - 2) < 2$.'

You need to find a 'range of x' — not just one value. Sounds a bit complicated — but it's not too bad once you get going. The first thing to do is rearrange the equation so you've got zero on one side.

Rearrange the expression to get...

$$(x+3)(x-2)<2$$
$$x^2+x-6<2$$
$$x^2+x-8<0$$

Rearrange this to get zero on one side. Then when you draw the graph, all you need to do is find where the graph is negative.

Then sketch the graph of $y = x^2 + x - 8$. And since you're interested in when this is less than zero, make sure you find out where this crosses the x-axis.

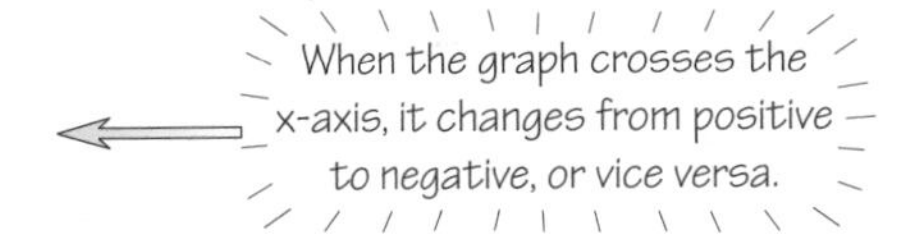

Now $x^2 + x - 8$ doesn't factorise — so find out where it crosses the x-axis by using the quadratic formula.

You can tell it doesn't factorise because $\sqrt{b^2 - 4ac} = \sqrt{33} = 5.74456...$ — and that's not a whole number or an 'easy' decimal.

Now, $x^2 + x - 8 = 0$ when

$$x = \frac{-1 \pm \sqrt{1^2 - (4 \times 1 \times -8)}}{2 \times 1}$$
$$= \frac{-1 \pm \sqrt{33}}{2}$$

The quadratic formula:

$$x = \frac{-b \pm \sqrt{b^2 - 4ac}}{2a}$$

when

$$ax^2 + bx + c = 0$$

So the graph looks like this:

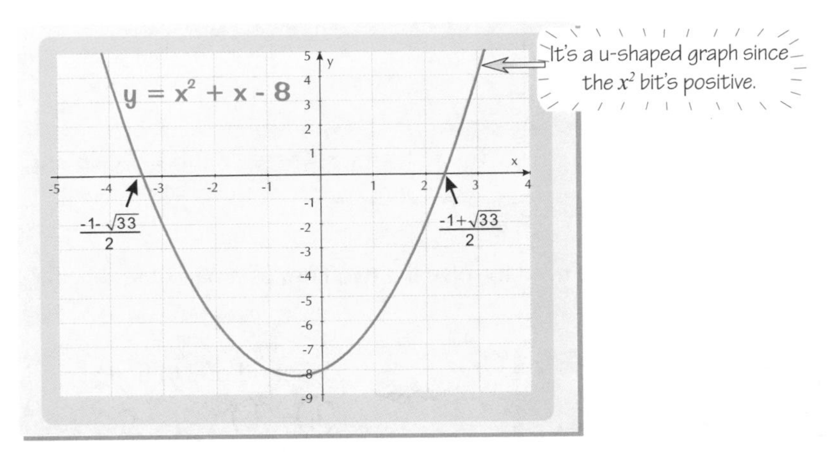

And since you need this to be negative — it's pretty clear that the range of x you're interested in is...

$$\frac{-1-\sqrt{33}}{2} < x < \frac{-1+\sqrt{33}}{2}$$

$a < x < b$ means 'x between a and b'.

or... $-3.37 < x < 2.37$ (to 2 d.p.)

It's easy to check — just stick the two numbers back into the original inequality and make sure the left-hand side equals 2.

These can be pretty darn hard unless you draw the graphs...

It's true. These questions can be very hard unless you have a picture to look at. It just helps you understand exactly what's going on, and what the examiners are going on about. That's the thing with these questions — they can look so intimidating. But drawing a picture helps you get your head round it — and once you've got your head round it, it's much easier to work towards the answer. Mmmm... I think I've said enough on that for now.

Paper 1 Q3 — Geometry

3 **(i)** Find the coordinates of the point A, when A lies at the intersection of the lines l_1 and l_2, and when the equations of l_1 and l_2 respectively are:

$$x - y + 1 = 0 \text{ and } 2x + y - 8 = 0.$$ [3]

(ii) The points B and C have coordinates (6, –4) and $(-\frac{4}{3}, -\frac{1}{3})$ respectively, and D is the midpoint of AC.

Find the equation of the line BD in the form $ax + by + c = 0$, where a, b and c are integers. [6]

(iii) Show that the triangle ABD is a right-angled triangle. [3]

(i) Finding A is easy — it's just Simultaneous Equations...

'Find the coordinates of the point A, when A lies at the intersection of the lines l_1 and l_2, and when the equations of l_1 and l_2 respectively are: $x - y + 1 = 0$ and $2x + y - 8 = 0$.'

Best draw a quick sketch so you know what you're looking for:

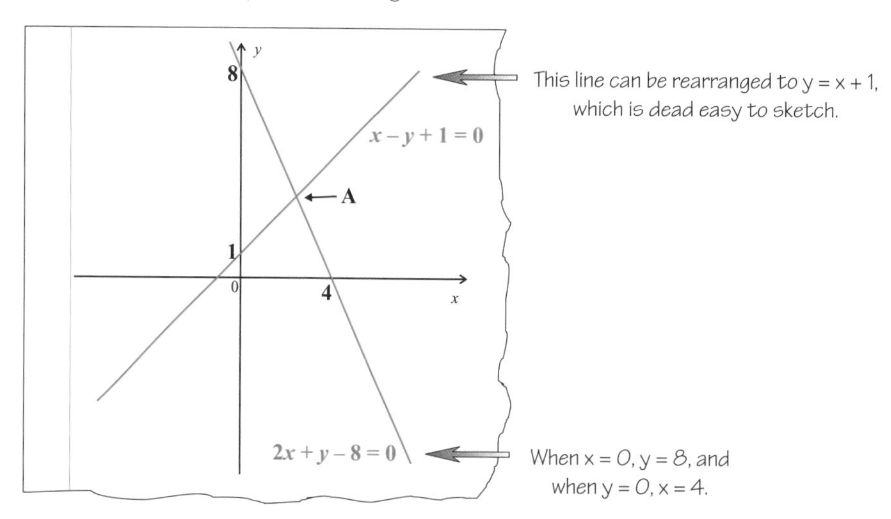

To find the coordinates of A, you need to solve the two lines as a pair of simultaneous equations.

Line 1 — (l_1) — $x - y + 1 = 0$

Line 2 — (l_2) — $2x + y - 8 = 0$

Get rid of y to find x:

(l_1) + (l_2)

$$(x + 2x) + (-y + y) + (1 - 8) = 0$$
$$3x - 7 = 0$$
$$x = \frac{7}{3}$$

Stick x = 7/3 back into l_1 to find y:

(l_1)

$$x - y + 1 = 0$$
$$\frac{7}{3} - y + 1 = 0$$
$$y = \frac{7}{3} + 1 = \frac{10}{3}$$

So A is: $\left(\frac{7}{3}, \frac{10}{3}\right)$

Forgotten everything you ever knew about simultaneous equations? Have a look at page 22.

Paper 1 Q3 — Geometry

(ii) Equation of a line — find the Gradient First...

'Find the equation of the line BD in the form $ax + by + c = 0$...'

This question gives you loads of information. So draw a sketch. Otherwise you won't have a clue what's going on. (Well I wouldn't anyway.)

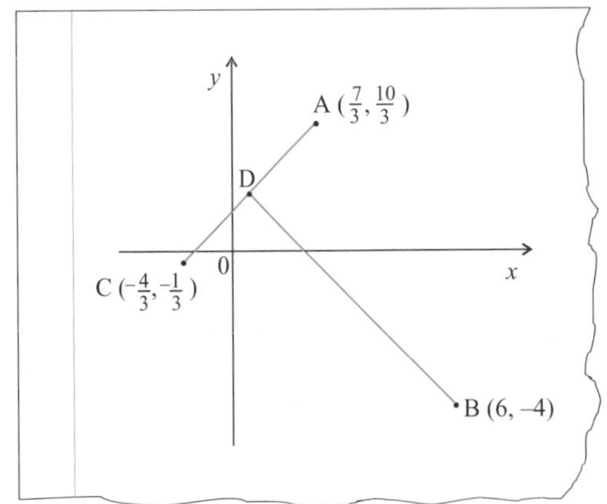

To find the equation of a line, you'll need its gradient.

And to find the gradient of BD, you'll need the coordinates of B and D — you're given B in the question, but you've got to find D.

Find D (the midpoint of A and C)

You get the midpoint of two points by finding the average of the x-coordinates, and the average of the y-coordinates.

Midpoint of AC is... $\left(\frac{x_A + x_C}{2}, \frac{y_A + y_C}{2}\right)$

x_A is the x-coordinate of the point A. y_C is the y-coordinate of the point C.

which is... $= \left(\frac{\frac{7}{3} + \frac{-4}{3}}{2}, \frac{\frac{10}{3} + \frac{-1}{3}}{2}\right) = (\frac{1}{2}, \frac{3}{2})$

So D is... $(\frac{1}{2}, \frac{3}{2})$

Find the Gradient of BD

$$\text{Gradient} = \frac{\text{difference in y-coordinates}}{\text{difference in x-coordinates}}$$

m_{BD} is the gradient of the line BD.

$$m_{BD} = \frac{y_D - y_B}{x_D - x_B} = \frac{\frac{3}{2} - -4}{\frac{1}{2} - 6} = \frac{3+8}{1-12} = -1$$

So you've got the gradient... Well now you can do anything — you can sail around the world, you can become the richest person in the world, you can rule the world... you can become more powerful than you can possibly imagine...

Find the Equation of BD

I reckon y = mx + c is the nicest form for the equation of a straight line. So I'd get it in that form first.

$$y = m_{BD}x + c \Rightarrow y = -x + c$$

The equation will be like this because we've just worked out that the gradient is –1 — you just need to find what c is.

Putting in the values for x and y at either point B or point D will give you the value of c.

At point B, x=6 and y=-4

$$y = mx + c$$
$$-4 = -6 + c$$
$$c = 2$$

So equation for BD is: $y = -x + 2$

$$x + y - 2 = 0$$

Make sure it's in the form the question asks for.

y = mx + c is great — m is the gradient, and c is where the line crosses the y-axis.

Paper 1 Q3 — Geometry

(iii) Right-angled triangle? Check if the lines are Perpendicular

'Show that the triangle ABD is a right-angled triangle.'

The first thing to do is update your sketch (or do a new one) with the triangle ABD on.

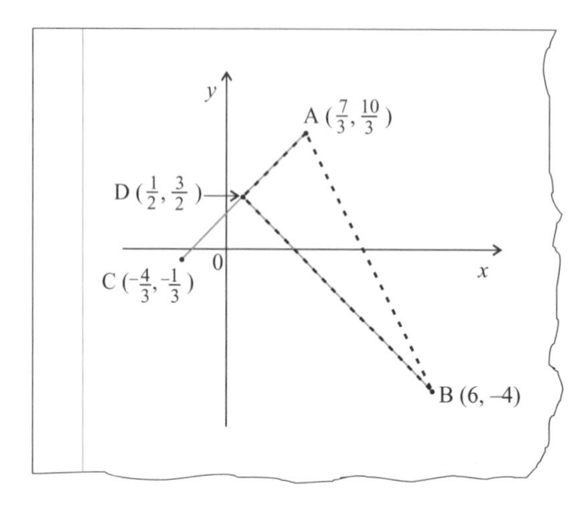

You've got to show that triangle ABD is right-angled. To do this, just show that two sides of the triangle are perpendicular to each other.
(That means you'll have a right angle in the corner where the sides meet.)

Hmm, what was that rule I used to know about perpendicular lines... ah yes, I remember:

Gradients of perpendicular lines multiply together to make –1.

See page 27 for more info on perpendicular lines.

Your sketch should show you that the right angle will be at D, so you need to show that the lines **AD** and **BD** are perpendicular.

Gradient of AD: $m_{DA} = \frac{y_D - y_A}{x_D - x_A}$

$= \frac{\frac{3}{2} - \frac{10}{3}}{\frac{1}{2} - \frac{7}{3}}$

$= \frac{9-20}{3-14} = 1$

And you already know the gradient of BD is: $m_{BD} = -1$

So... $m_{BD} \times m_{DA} = -1 \times 1 = -1$

So angle ADB is a right angle. Fantastic.

So you've proved it's a right-angled triangle and completed a stinker of a question. Now you can...

...become the Master of the Universe...

To avoid getting into trouble while you're doing this question, you've got to draw what's happening before you do each part. So when you read stuff like, "The points B and C have coordinates (6, –4) and...", you should dive for your pencil and sketch all the information it gives you. And then use the sketch and plan how you're going to answer the question.

Paper 1 Q4 — Simultaneous Equations

4 Solve the following simultaneous equations:

$y + x = 7$ and $y = x^2 + 3x - 5$ [4]

Some Simultaneous Equations to solve, but one's a Quadratic...

'Solve the simultaneous equations: $y + x = 7$ and $y = x^2 + 3x - 5$.'

1) When you see simultaneous equations with a quadratic, you need the **substitution** method.
2) You need to **rearrange** the linear equation to get x or y on its **own**.
3) You can do it for either x or y, but the working's easier if you substitute for y.

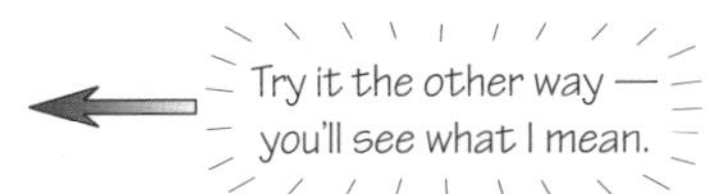

The linear equation: $y + x = 7$

Rearrange: $y = 7 - x$

Substitute that into the quadratic: $y = x^2 + 3x - 5$

which becomes: $7 - x = x^2 + 3x - 5$

Rearrange again to get everything on one side of the equation: $0 = x^2 + 4x - 12$

which you can of course factorise: $(x + 6)(x - 2) = 0$

You remember how to factorise, don't you? Just put your lips together and blow. Well, actually — there's a bit more to it than that...

Write out brackets with the xs in: $x^2 + 4x - 12 = (x \quad)(x \quad)$

Then try pairs of numbers that multiply together to make −12: $(x - 12)(x + 1)$, $(x + 12)(x - 1)$, $(x + 6)(x - 2)$ etc.

Then pick one where the numbers add together to make +4. It's got to be: $(x + 6)(x - 2)$

because: $+6 - 2 = +4$ Grand.

So the two values of x you want are: $x = -6$ and $x = 2$

Don't fall into the trap of getting this far and thinking you've finished the question.
You need the **corresponding values of *y*** too. Just stick the two values of x back into the linear equation.

$y + x = 7$

When $x = -6$:
$y + -6 = 7$
$y = 13$

When $x = 2$:
$y + 2 = 7$
$y = 5$

So the two solutions are: $(-6, 13)$ and $(2, 5)$.

Now check these answers work in both equations:

$(-6, 13)$
$y + x = 7$
$13 + -6 = 7$ ✓
$y = x^2 + 3x - 5$
$13 = (-6)^2 + (3 \times -6) - 5$
$13 = 36 - 18 - 5$ ✓

$(2, 5)$
$y + x = 7$
$5 + 2 = 7$ ✓
$y = x^2 + 3x - 5$
$5 = 2^2 + (3 \times 2) - 5$
$5 = 4 + 6 - 5$ ✓

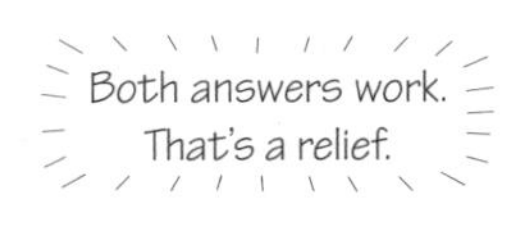

"Substitute your lies for fact" — good advice there from The Who...*

There's nothing massively hard with this kind of question — you substitute, and then get on with solving stuff. But there's always the temptation to stop after you've solved the quadratic. This is a bad move and upsets examiners no end — remember, they'll think you're nicer if all your answers have an x- and a y-value.

*(Although a lot of people maybe don't know who The Who are, thinking about it.)

Paper 1 Q5 — Quadratics

5 **(i)** Express $x^2 - 6x + 5$ in the form $(x + a)^2 + b$. [2]

(ii) Factorise the expression $x^2 - 6x + 5$. [2]

(iii) Hence sketch the graph of $y = x^2 - 6x + 5$, clearly indicating the coordinates of the vertex and the points where it cuts the axes. [3]

(i) Complete the Square

'Express $x^2 - 6x + 5$ in the form $(x + a)^2 + b$.'

Always start by trying to find a because it's dead easy. You just divide the number in front of the x by two.

In this case, half of –6 is –3: $x^2 - 6x + 5 = (x - 3)^2 + b$ ($\div 2$)

Now multiply out... $(x - 3)^2 + b = x^2 - 6x + 9 + b$

then equate to find b: $x^2 - 6x + 5 = x^2 - 6x + 9 + b$

So: $5 = 9 + b$

$b = -4$

...then write it all out neatly — ta daaa:

$x^2 - 6x + 5 = (x - 3)^2 - 4$

(ii) Oh Hallelujah — it's another Quadratic...

'Factorise the expression $x^2 - 6x + 5$.'

Write out the two brackets — the coefficient of x^2 is 1, so start by writing x at the front of each bracket.

$x^2 - 6x + 5 = (x \quad)(x \quad)$

The numbers in the brackets have to multiply to make 5 and add together to make –6.

Oh look... –1 and –5 might just do it... $-1 + -5 = -6$ and $-1 \times -5 = 5$

$x^2 - 6x + 5 = (x - 1)(x - 5)$ Hurrah.

Quadratics are nice'n'easy to factorise when the number at the end is prime, because it only has one pair of factors.

(iii) Find the Key Points Before you Sketch

'Hence sketch the graph of $y = x^2 - 6x + 5$, clearly indicating the coordinates of the vertex and the points where it cuts the axes.'

"Hence" means you can use the work you've already done.

In part (i) you showed that $x^2 - 6x + 5 = (x - 3)^2 - 4$. You can use this to find the vertex.

$(x - 3)^2$ can't be less than zero (because it's squared, of course).

So the minimum must be when the bracket = 0, i.e. when: $x = 3$

and therefore when: $y = 0 - 4 = -4$

So the vertex of the graph is at (3, –4).

The coefficient of x^2 is positive, so the graph is U-shaped, and you're looking for a minimum rather than a maximum.

Paper 1 Q5 — Quadratics

Now find where the graph crosses the two axes:

When the curve cuts the x-axis, y equals zero. So: $x^2 - 6x + 5 = 0$

Using your result from part (ii), you have: $(x - 1)(x - 5) = 0$

So $x = 1$ and $x = 5$, i.e. the graph cuts the x-axis at: $(1, 0)$ and $(5, 0)$

Two solutions, so your graph crosses the x-axis twice.

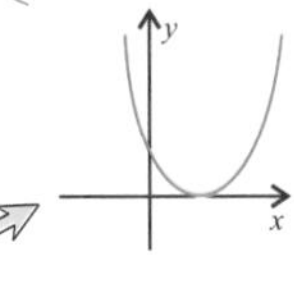

Some quadratics meet the x-axis just once, while others never have the pleasure.

When the curve cuts the y-axis, x equals zero. So: $y = 0^2 - (6 \times 0) + 5$

This gives: $y = 5$

So the graph cuts the y-axis at: $(0, 5)$

One solution, so your graph crosses the y-axis once. This is actually the case for all quadratics, but sometimes you can't see it happen.

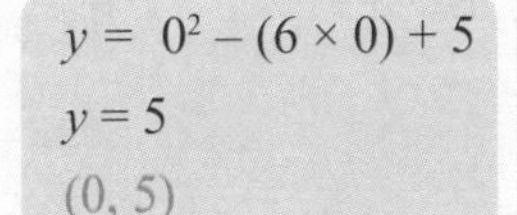

They'll meet eventually...

Now you're ready for some Sketching Action

You know that a quadratic graph with a positive coefficient of x^2 will be a u-shape.
You now know the key points too, so plotting the graph is easy.

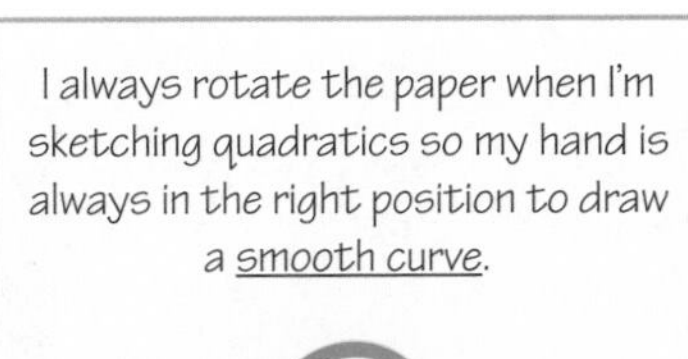

I always rotate the paper when I'm sketching quadratics so my hand is always in the right position to draw a smooth curve.

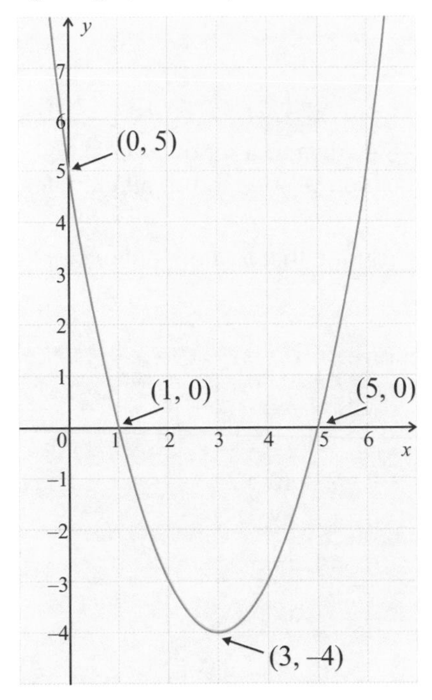

Knowing where the graph crosses the axes will help you decide the scale of the sketch.

Make sure you give yourself plenty of room to plot and label all the key points.

Another thing: check the entire curve is perfectly symmetrical about a vertical line that goes through the vertex. For example, you know the curve goes through (0, 5), so it must go through (6, 5) on the opposite side.

If "quad" means "4" — why are quadratics called quadratics...

What can I say about quadratics that hasn't been said already... probably nothing, so let me just repeat something that you've heard before. Quadratics questions should be meat and drink. When you see the word quadratics, think factorising, think completing the square, think quadratic formula, think sketching parabolas. If you think those 4 things, chances are one of them will work on the question in hand.

Paper 1 Q6 — Factorising and Integrating

6 A curve that passes through the point (2, 0) has derivative $dy/dx = 3x^2 + 6x - 4$.

(i) Show that the equation of the curve is $y = x^3 + 3x^2 - 4x - 12$. [3]

(ii) $(x + 3)$ is a factor of y. Express y as a product of 3 linear factors. [2]

(i) Integrate dy/dx to find y

'A curve that passes through the point (2, 0) has derivative $dy/dx = 3x^2 + 6x - 4$.

i) Show that the equation of the curve is $y = x^3 + 3x^2 - 4x - 12$.'

When you see dy/dx, it's easy to think "aha — differentiation!".
But in this question, you're told dy/dx and you have to work **backwards** to find y.
So you have to do the **opposite** and **integrate**.

$$\frac{dy}{dx} = 3x^2 + 6x - 4$$

Integrate each side:

$$y = \frac{3x^3}{3} + \frac{6x^2}{2} - 4x + c$$

$$y = x^3 + 3x^2 - 4x + c$$

Don't forget the constant when you integrate.

With these 'Show that...' questions, it's really important to write down your method clearly — it's what gets you the marks.

The question gives you 2 bits of info — dy/dx and a set of coordinates.
If you're not sure what to do, make sure you've tried to use all the info you've been given.
The curve has to go through (2, 0), so you use that to work out what c is.

Just put $x = 2$ and $y = 0$ into the equation: $y = x^3 + 3x^2 - 4x + c$.

when $x = 2$ and $y = 0$:

$$0 = 2^3 + (3 \times 2^2) - (4 \times 2) + c$$

$$0 = 8 + 12 - 8 + c$$

$$0 = 12 + c$$

so $c = -12$,

which means your equation is: $y = x^3 + 3x^2 - 4x - 12$ — which is exactly what you want.

Usually you should check your answer by differentiating to make sure you get back to where you started — but here you don't need to, because the question tells you the answer already. You're just showing that you get the same equation.

Paper 1 Q6 — Factorising and Integrating

(ii) Now Factorise Completely

'$(x + 3)$ is a factor of y. Express y as a product of 3 linear factors.'

You know one factor of $x^3 + 3x^2 - 4x - 12$ is $(x + 3)$. To factorise completely, you have to find what you multiply $(x + 3)$ by to get $x^3 + 3x^2 - 4x - 12$.

$$(x + 3)(\ldots\ldots\ldots\ldots) = x^3 + 3x^2 - 4x - 12$$

This bracket will be quadratic (to give the 'x^3' on the RHS). The first and last terms are easy to find:

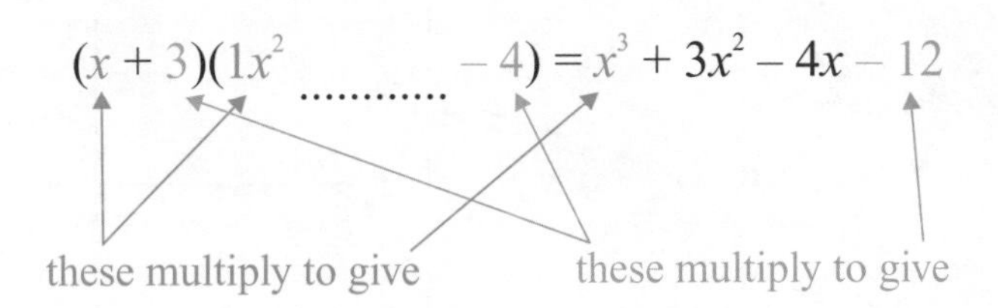

You can see $3 \times 1x^2$ gives the right number of x^2s and $x \times (-4)$ gives the right number of xs ...so you've actually got all you need for the second bracket.

$$(x + 3)(x^2 - 4) = x^3 + 3x^2 - 4x - 12$$

Now you need to get two factors out of $(x^2 - 4)$. Any ideas?
You need to spot that it's a **difference of two squares**.

$$(x^2 - 4) = (x + 2)(x - 2)$$

You can now write y as the product of 3 linear factors:

$$y = x^3 + 3x^2 - 4x - 12 = \mathbf{(x + 3)(x + 2)(x - 2)}$$

Rather elegant, I'm sure you'll agree. The algebraic equivalent of a mink coat, only slightly less controversial.

Calculus and algebra in one question — Fiendish...

I know it seems like these questions go on for ever, but I'm sure you can do all the individual bits — it's just a question of stringing it all together. But notice how the question leads you through. When it says "Show...", at least you know what you're aiming for. Aren't they good to you, those examiners... (Actually, don't answer that.)

Paper 1 Q7 — Equation of a Line

7 The line AB is part of the line with equation $y + 2x - 5 = 0$.
A is the point with coordinates (1, 3) and B is the point with coordinates (4, k).

(i) Find the value of k. [1]

(ii) What is the equation of the line perpendicular to AB, that passes through A? [3]

(i) Do a Sketch and a bit of Substitution

'Find the value of k.'

Always, always, always do a sketch if you get a question like this.
No ifs, no buts — a sketch will always help.
It's a straight-line graph so it's pretty easy.
Just work out where it crosses the axes:

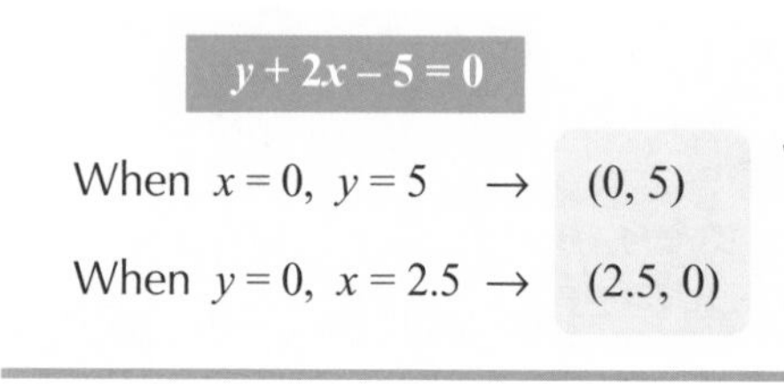

$y + 2x - 5 = 0$

When $x = 0$, $y = 5$ → (0, 5)

When $y = 0$, $x = 2.5$ → (2.5, 0)

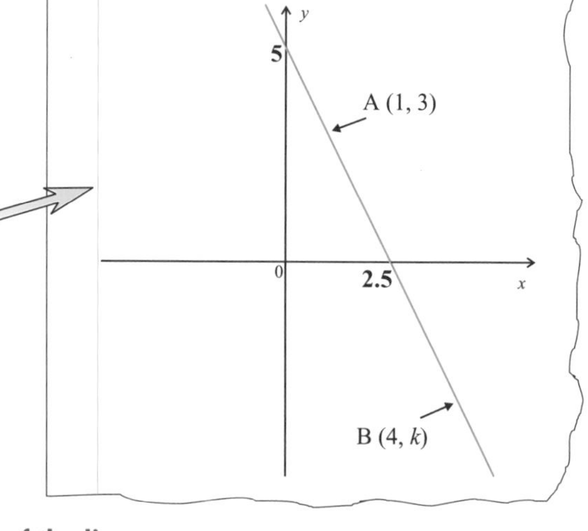

The sketch shows that you're looking for a negative value of k.
You can find the exact value by **substituting $x = 4$ into the equation of the line**:

$y + 2x - 5 = 0$

When $x = 4$: $y + (2 \times 4) - 5 = 0$

$y + 8 - 5 = 0$

$y + 3 = 0$

$y = -3$

So: $k = -3$

It's so easy to get these questions wrong. If you've drawn a sketch, you'll know whether or not your answer is **sensible**.

(ii) Use the Gradient Rule for Perpendicular Lines

'What is the equation of the line perpendicular to AB that passes through A?'

For this part you need to remember something about the gradients of perpendicular lines. Ring any bells?
If you didn't know the gradient rule already, make sure you LEARN IT:

The gradients of two perpendicular lines multiply together to make –1.

This means that the gradient of your new line will be: $\frac{-1}{\text{the gradient of AB}}$

Paper 1 Q7 — Equation of a Line

To find the gradient of AB, you need the equation in the form: $y = mx + c$.

Time for a bit of rearranging: $y = -2x + 5$ Lo and behold, the gradient of AB is: -2

So the gradient of the perpendicular line is: $\frac{-1}{-2} = \frac{1}{2}$

Still with me ...

To get the equation of the line, use:

$$y - y_1 = m(x - x_1)$$
$$y - 3 = \tfrac{1}{2}(x - 1)$$
$$y - 3 = \tfrac{1}{2}x - \tfrac{1}{2}$$
$$y = \tfrac{1}{2}x + 2\tfrac{1}{2}$$

"m" is the gradient which you just found, and (x_1, y_1) are the coordinates of a point on the line — in this case A (1, 3).

Here's another nifty way of finding perpendicular lines:

1) Swap the numbers in front of x and y round, and then
2) Make one of them negative (forget about the number term for now): $2y - x + ? = 0$
3) Use the point you know on the line (A) to work out the number bit. Subbing in (1, 3) gives: $(2 \times 3) - 1 + ? = 0$
4) So the missing number is –5 and the equation is: $2y - x - 5 = 0$

(If you add $x + 5$ to both sides: $2y = x + 5$ and then divide both sides by 2: $y = \tfrac{1}{2}x + 2\tfrac{1}{2}$ which is the same answer as the other method.)

Whichever method you've used, remember to...

...Sketch it to Make Sure

Go back to your masterpiece from part (i).
You want to check that the new line goes through A, and that the two lines are perpendicular.

Same drill — find where the line crosses the axes.

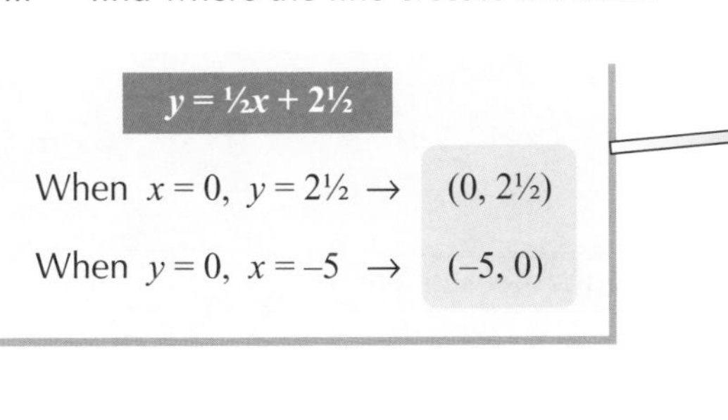

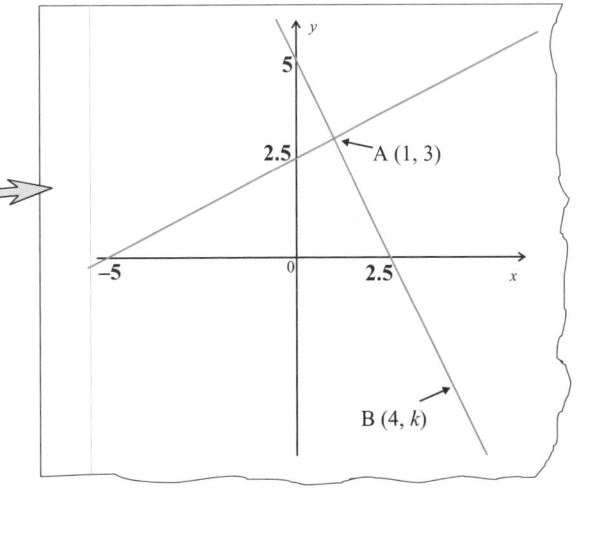

You don't need to use a protractor to check the two lines are at right angles. Just make sure the lines look about right (e.g. not like this: ><)

Looks good to me.

Pretty nifty, that gradient rule...

...and very easy to learn. Anyway, all this should be bread-and-butter stuff to you by now. If so, go and make yourself a sandwich (boom boom). If not, go back over this question and make sure you can do it from start to finish. Just imagine going into the exam hall, finding your desk, turning over the paper and then seeing a question just like this one. How gutted will you feel if you can't do it, and you have only a vague memory of something similar you've seen somewhere before...

Paper 1 Q8 — Arithmetic Series

8 (i) An arithmetic series has first term a and common difference d.

(a) Write down expressions for u_n, u_{n+1} and u_{n+2}, the nth, $(n+1)^{th}$ and $(n+2)^{th}$ terms in the series respectively. [2]

(b) By making the substitution $x = a + nd$, write these expressions in terms of x and d only. [4]

(c) If the sum of u_n, u_{n+1} and u_{n+2} is 36, and their product is 960, find the positive values of x and d. [4]

(d) If u_n, u_{n+1} and u_{n+2} above are the first three terms of the series, write down an expression for u_n in terms of n. [3]

(ii) Find S_{10}, the sum of the first ten terms of the series. [3]

(iii) By considering the formula for S_n and the formula for the sum $\sigma_n = 1 + 2 + 3 + ... + n$, find an expression for the difference $S_n - \sigma_n$, giving your answer in as simple a form as possible. [4]

(i)(a) The nth Term of an Arithmetic Series — piece of cake

'Write down expressions for u_n, u_{n+1} and u_{n+2}...'

The question uses the words 'write down' (not 'work out' or 'find', or anything like that) — so you should just be able to quote the answer without thinking about it.

For an arithmetic series, the nth term is just: $u_n = a + (n - 1)d$, so...

$$u_n = a + (n-1)d$$
$$u_{n+1} = a + nd$$
$$u_{n+2} = a + (n+1)d$$

Use the same formula each time, but substitute (n+1) and (n+2) for n.
E.g. $u_{n+1} = a + [(n + 1) - 1]d = a + nd$, etc.

(i)(b) Substitute — and then Rewrite

'By making the substitution $x = a + nd$, write these expressions in terms of x and d only.'

So you've got to rewrite those same expressions, but with only x and d on the right-hand side. But at least it tells you how to do it — by writing x instead of '$a + nd$'.

$$u_n = a + (n-1)d = a + nd - d = x - d$$
$$u_{n+1} = a + nd = x$$
$$u_{n+2} = a + (n+1)d = a + nd + d = x + d$$

Work out the brackets in the answers to part (i), and then write x wherever you see 'a+nd'.

Paper 1 Q8 — Arithmetic Series

(i)(c) Fiddle About a bit to find x and d

'If the sum of u_n, u_{n+1} and u_{n+2} is 36, and their product is 960, find the positive values of x and d.'

This looks hard — but there are some big clues in the question. It tells you what the sum and product of the expressions in part (b) equal — so start by working those out...

The sum of a load of numbers is what you get when you add them all together.
The product of a load of numbers is what you get when you multiply them all together.

The sum is:

$$u_n + u_{n+1} + u_{n+2} = (x-d) + x + (x+d)$$
$$= 3x$$

And this equals 36, so...

$$3x = 36$$
$$\Rightarrow x = 12$$

The product is:

$$u_n \times u_{n+1} \times u_{n+2} = (x-d)x(x+d)$$
$$= x(x^2 - d^2)$$

Using the difference of two squares: $(x + d)(x - d) = x^2 - d^2$.

This equals 960. But you've just worked out that $x = 12$ — so put this value in as well...

$$x(x^2 - d^2) = 960$$
$$\Rightarrow 12(144 - d^2) = 960$$

This only has one letter left to find, d.

And now you can find the value of d...

$\frac{960}{12} = 80$

$$\Rightarrow 144 - d^2 = 80$$
$$\Rightarrow d^2 = 144 - 80 = 64$$
$$\Rightarrow d = 8$$

(i)(d) Now find the First Term using your answers from before...

'...write down an expression for u_n in terms of n.'

If u_n, u_{n+1} and u_{n+2} are the first three terms of the series, then $u_1 = u_n = x - d = 12 - 8 = 4$

Using your answers from parts (b) and (c).

...which means the nth term of the series is:

$$u_n = a + (n-1)d = 4 + 8(n-1)$$
$$= 4 + 8n - 8 = 8n - 4$$

You've just worked out that the first term of the series u_1 (= a) is 4.

Paper 1 Q8 — Arithmetic Series

(ii) Sum an arithmetic series

'Find S_{10}, the sum of the first ten terms of the series.'

This is an absolute doddle — as long as you know the formula ...

The formula for the sum of the first n terms of an arithmetic series is:

$$S_n = \frac{n}{2}[2a+(n-1)d]$$

This is the general formula for the sum of the first n terms...

And so if $n = 10$: $S_{10} = 5[2a+9d]$

...so if you only want the sum of the first ten terms, use this one.

Or you can use the other formula for the sum of the first n terms of an arithmetic series: $S_n = \frac{n}{2}(a+l)$.

Putting in the values of a (= 4) and d (= 8), you get...

$$S_{10} = 5[(2\times4)+(9\times8)]$$
$$= 5\times80 = 400$$

(iii) What do the examiners want now — Blood...

'...find an expression for the difference $S_n - \sigma_n$...'

Right then — now it's serious. This bit looks nasty — but just do it one bit at a time.

You've already got the formula for S_n. It is:

$$S_n = \frac{n}{2}[2a+(n-1)d]$$

And since you know $a = 4$ and $d = 8$, this becomes:

$$S_n = \frac{n}{2}[8+8(n-1)]$$
$$= \frac{n}{2}[8+8n-8]$$
$$= \frac{n}{2}(8n) = 4n^2$$

Now $\sigma_n = 1 + 2 + 3 + ... + n$, and you should know that this sum is given by:

$$\sigma_n = 1+2+...+n = \sum_{i=1}^{n} i = \frac{n}{2}(n+1)$$

You'll need to remember this formula.

And all the question's actually asking you to do is work out $S_n - \sigma_n$. This is just:

$$S_n - \sigma_n = 4n^2 - \frac{n}{2}(n+1)$$
$$= 4n^2 - \frac{n^2}{2} - \frac{n}{2}$$
$$= \frac{7}{2}n^2 - \frac{n}{2}$$
$$= \frac{n}{2}(7n-1)$$

Taking a factor of $\frac{n}{2}$ outside the brackets.

Check your answer: choose n = 3.
$S_3 = 4 + 12 + 20 = 36$.
$\sigma_3 = 1 + 2 + 3 = 6$.
According to your formula the difference between them should be: $\frac{3}{2}((7\times3)-1) = 30$. And since $36 - 6 = 30$, this works.

When the question says jump — you ask, 'How high...'

A lot of questions look mean, but when you get past the 'maths-speak' they're not really so bad. When they give 'advice' on tackling a problem, it's a good idea to take it — do as you're told, basically. Also, with this question (like a load of others), you can check your answer by choosing a (small) value for n and seeing if your formula works. Then if it does work, you can do the next question with a warm happy feeling inside. Wonderful...

Paper 1 Q9 — Differentiation

9 Find dy/dx for each of the following:

(i) $y = x^2$ [1]

(ii) $y = 3x^4 - 2x$ [2]

(iii) $y = (x^2 + 4)(x - 2)$ [2]

(i) Just use the Rule for Differentiation

'Find dy/dx for $y = x^2$.'

When you see dy/dx or $f'(x)$, it means you need to **differentiate**.

If you've forgotten the rule for differentiating powers of x, flick back to page 35.

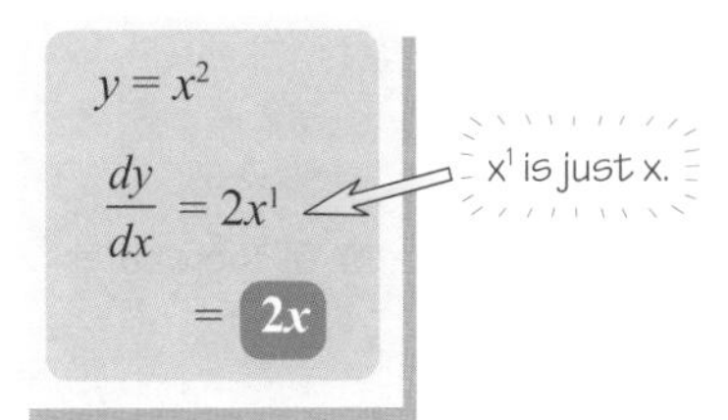

(ii) Apply the rule to Each Term in Turn

'Find dy/dx for $y = 3x^4 - 2x$.'

Again you can use the rule for differentiation — but there are two terms, so you do it twice.

For the first term $n = 4$, and for the second term $n = 1$:

$$y = 3x^4 - 2x$$

$$\frac{dy}{dx} = 3(4x^{4-1}) - 2(1x^{1-1})$$

$$= 3(4x^3) - 2(1)$$

$$= 12x^3 - 2$$

Don't let the coefficients in front of the x bits put you off — you just multiply by them. In maths-speak, that's $d/dx\,(ky) = k\,dy/dx$.

(iii) Expand first — then Differentiate

'Find dy/dx for $y = (x^2 + 4)(x - 2)$.'

When you need to differentiate an expression like this, you have to multiply it out first.

$$y = (x^2 + 4)(x - 2)$$

$$y = x^3 - 2x^2 + 4x - 8$$

This gives you a set of terms which are all powers of x. Which makes everything lovely.
Now differentiate each term:

$$y = x^3 - 2x^2 + 4x - 8$$

$$\frac{dy}{dx} = 3x^{3-1} - 2(2x^{2-1}) + 4(1x^{1-1}) - 0$$

$$= 3x^2 - 4x + 4$$

Constant terms always disappear when you differentiate. That's because for a constant, the power of x is 0, so when you differentiate, you end up multiplying the term by 0, which gives 0. Glad we've cleared that up...

It'd be great if the summer term disappeared...

...and the spring one and the autumn one. Differentiation strikes fear into the hearts of the hardest of maths students. It can get a little sticky at times, so make sure you know the rules really well.
Even then you probably won't be laughing, but at least you might be slightly more optimistic.

General Certificate of Education
Advanced Subsidiary (AS) and Advanced Level

Core 1 Mathematics — Practice Exam Two

Give non-exact numerical answers correct to 3 significant figures, unless a different degree of accuracy is specified in the question or is clearly appropriate.

1 **(i)** Express $x^2 - 7x + 17$ in the form $(x-a)^2 + b$, where a and b are constants.

Hence state the maximum value of $f(x) = \dfrac{1}{x^2 - 7x + 17}$. [3]

(ii) Find the possible values of b if the equation $g(x) = 0$ is to have only one root,

where $g(x)$ is given by $g(x) = 3x^2 + bx + 12$. [3]

2 **(i)** Simplify $(\sqrt{3}+1)(\sqrt{3}-1)$. [1]

(ii) Rationalise the denominator of the expression $\dfrac{\sqrt{3}}{\sqrt{3}+1}$. [2]

3 A triangle has sides which lie on the lines given by the following equations:

AB: $y = 3$ BC: $2x - 3y - 21 = 0$ AC: $3x + 2y - 12 = 0$

(i) Find the coordinates of the vertices of the triangle. [5]

(ii) Show that the triangle is right-angled. [2]

The point D has coordinates $(3, d)$ and the point E has coordinates $(9, e)$.

(iii) If point D lies outside the triangle, but not on it, show that either $d > 3$ or $d < 1.5$. [2]

(iv) Given that E lies inside the triangle, but not on it, find the set of possible values for e. [3]

4 A curve has equation $y = f(x)$, where $dy/dx = 4(1 - x)$.

The curve passes through the point A, with coordinates (2,6).

(i) Find the equation of the curve. [4]

(ii) Sketch the graph of $y = f(x)$. [3]

5 **(i)** Solve the equation: $\dfrac{4}{(x-2)} = \dfrac{6}{(2x+5)}$. [3]

(ii) Show that: $\dfrac{4}{9(x+2)} + \dfrac{1}{3(x-1)^2} + \dfrac{5}{9(x-1)} = \dfrac{x^2}{(x+2)(x-1)^2}$. [5]

6 A curve has the equation $y = f(x)$, where $f(x) = x^3 - 3x + 2$.

(i) Find dy/dx. [2]

(ii) Find the points at which the gradient of the curve $y = f(x)$ is 0. [4]

(iii) Show that $x^3 - 3x + 2$ factorises to $(x-1)^2(x+2)$. [2]

(iv) Sketch the graphs of:

(a) $y = f(x)$ [3]

(b) $y = f(x - 3)$ [2]

(c) $y = 2f(x)$ [2]

7 A function is defined by $f(x) = x^3 - 4x^2 - 7x + 10$.

$(x - 1)$ is a factor of $f(x)$.

Hence or otherwise solve the equation $x^3 - 4x^2 - 7x + 10 = 0$. [2]

8 **(i)** What is the exact value of 5^{-2}? [1]

(ii) What is the exact value of $(4^{\frac{1}{2}})^6$? [1]

(iii) Simplify the expression: $\dfrac{x^3 \times x^4}{\sqrt{x^{10}}}$ [2]

9 The diagram shows part of the graph with equation $y = x^3 - 2x^2 + 4$.

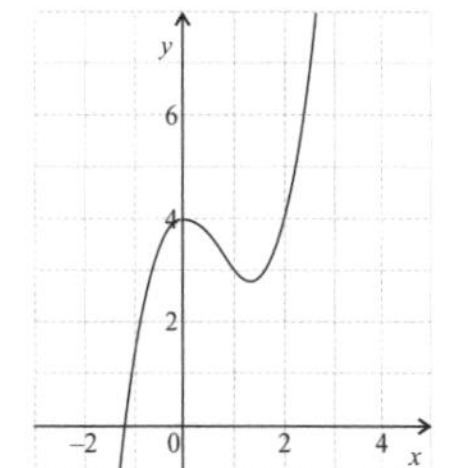

(i) Find the equation of the tangent at $x = 1$. [3]

(ii) Find the equation of the normal to the curve at $x = 2$. [3]

(iii) Find the distance between where the tangent and the normal cross the x-axis. [2]

10 **(i)** A sequence is defined by the recurrence relation $x_{n+1} = 3x_n - 4$.

Given that the first term is 6, find x_4. [2]

(ii) The third term of an arithmetic progression is 9 and the seventh term is 33.

(a) Find the first term and the common difference. [3]

(b) Find S_{12}, the sum of the first 12 terms in the series. [3]

(c) Hence or otherwise find: $\sum_{1}^{12}(6n+1)$ [2]

Paper 2 Q1 — Quadratics

1 **(i)** Express $x^2 - 7x + 17$ in the form $(x-a)^2 + b$, where a and b are constants.
Hence state the maximum value of $f(x) = \dfrac{1}{x^2 - 7x + 17}$. [3]

(ii) Find the possible values of b if the equation $g(x) = 0$ is to have only one root,
where $g(x)$ is given by $g(x) = 3x^2 + bx + 12$. [3]

(i) An easy Completing the Square bit

'Express $x^2 - 7x + 17$ in the form $(x-a)^2 + b$, where a and b are constants.'

The question's just asking you to complete the square. The x^2 bit's got a coefficient of 1, so it's not so bad...

$$x^2 - 7x + 17$$

Rewrite it as one bracket squared + a number.

$$= \left(x - \tfrac{7}{2}\right)^2 + b$$

Don't forget — you just halve the coefficient of x to get the number in the brackets.

Find b by making the old and new expressions equal to each other.

$$x^2 - 7x + 17 = \left(x - \tfrac{7}{2}\right)^2 + b$$

$$x^2 - 7x + 17 = x^2 - 7x + \frac{49}{4} + b$$

$$b = 17 - \frac{49}{4} = \frac{19}{4}$$

$$x^2 - 7x + 17 = \left(x - \tfrac{7}{2}\right)^2 + \frac{19}{4}$$

Simplify the equation to get the value of b.

Have a look at pages 12 and 13 for more on completing the square.

Finding the Maximum...

'Hence state the maximum value of $f(x) = \dfrac{1}{x^2 - 7x + 17}$.'

It'll be maximum when the quadratic in the denominator is as small as possible.

The minimum value of the denominator is easy to see using your answer to the first part — it's $\frac{19}{4}$, because the squared bit is never less than zero.

So... $f(x)_{max} = \dfrac{1}{\left(19/4\right)} = \dfrac{4}{19}$

It says 'hence' — that's a pretty major clue that you've got to use the result from the first part to do this bit.

(ii) Use the Discriminant

'Find the possible values of b if the equation $g(x) = 0$ is to have only one root, where $g(x) = 3x^2 + bx + 12$.'

You can tell how many roots a function has got if you use the quadratic formula.
Think about $b^2 - 4ac$. If $b^2 - 4ac = 0$, then it's only got one root.

$$x = \frac{-b \pm \sqrt{b^2 - 4ac}}{2a}$$

The b^2–4ac bit is called the discriminant — and it's this part that tells you how many roots a quadratic has. See page 15.

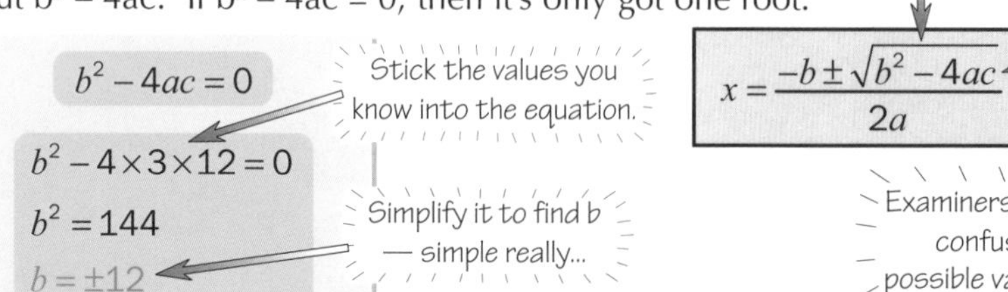

$$b^2 - 4ac = 0$$

$$b^2 - 4 \times 3 \times 12 = 0$$

$$b^2 = 144$$

$$b = \pm 12$$

Stick the values you know into the equation.

Simplify it to find b — simple really...

Examiners love adding things like g(x) — probably just to confuse you. They could have just said: 'Find the possible values of b if $3x^2 + bx + 12 = 0$ has only one root.'

The square is complete..... Now I am the Master...

One of the commonest completing-the-square mistakes is not noticing when there's a number in front of the x^2 bit. Either that or noticing it but trying to stick it inside the brackets. Uh-uh. You've got to stick it outside, so it multiplies both the x^2 and x parts. And with the quadratic formula — you've got to be oh, so careful with those minus signs. So watch it.

Paper 2 Q2 — Surds

2 **(i)** Simplify $(\sqrt{3}+1)(\sqrt{3}-1)$. [1]

(ii) Rationalise the denominator of the expression $\frac{\sqrt{3}}{\sqrt{3}+1}$. [2]

(i) Just Multiply it out

'Simplify $(\sqrt{3}+1)(\sqrt{3}-1)$'

Just multiply it out like you would if it was a quadratic:

$$(\sqrt{3}+1)(\sqrt{3}-1) = \sqrt{3}\times\sqrt{3} - 1\times\sqrt{3} + 1\times\sqrt{3} - 1\times 1$$
$$= 3 - \sqrt{3} + \sqrt{3} - 1$$
$$= 2$$

Rules of Surds

There's not really very much to remember.

$$\sqrt{ab} = \sqrt{a}\sqrt{b}$$
$$\sqrt{\frac{a}{b}} = \frac{\sqrt{a}}{\sqrt{b}}$$
$$a = (\sqrt{a})^2 = \sqrt{a}\sqrt{a}$$

Better still, you can save a bit of time if you recognise it as the **difference of 2 squares**: $(a+b)(a-b) = a^2 - b^2$

$$(\sqrt{3}+1)(\sqrt{3}-1) = (\sqrt{3})^2 - 1^2$$
$$= 3 - 1$$
$$= 2$$

(ii) Get rid of the Square Root in the Denominator

'Rationalise the denominator of the expression $\frac{\sqrt{3}}{\sqrt{3}+1}$.'

To get rid of a root from the bottom of a fraction, you use the difference of two squares — e.g. if the denominator's $\sqrt{a}+b$, you multiply by $\sqrt{a}-b$.

In this case, the denominator's $\sqrt{3}+1$, so you have to multiply by $\sqrt{3}-1$.

To get an equal fraction, you multiply the **top and bottom** of the fraction by the same amount.

$$\frac{\sqrt{3}}{\sqrt{3}+1} = \frac{\sqrt{3}(\sqrt{3}-1)}{(\sqrt{3}+1)(\sqrt{3}-1)} = \frac{3-\sqrt{3}}{2}$$

The bottom bit's just what you worked out for part (i).

To ignore surds would just be irrational...

Surds are really nice because they're clean — there are no messy decimals involved. They also save you the bother of using your calculator. Make sure you know the rules of surds shown at the top, and then get plenty of practice at applying these rules — things like rationalising the denominator are very likely to come up in the exam.

Paper 2 Q3 — Simultaneous Equations

3 A triangle has sides which lie on the lines given by the following equations:

AB: $y = 3$ BC: $2x - 3y - 21 = 0$ AC: $3x + 2y - 12 = 0$

(i) Find the coordinates of the vertices of the triangle. [5]

(ii) Show that the triangle is right-angled. [2]

The point D has coordinates $(3, d)$ and the point E has coordinates $(9, e)$.

(iii) If point D lies outside the triangle, but not on it, show that either $d > 3$ or $d < 1.5$. [2]

(iv) Given that E lies inside the triangle, but not on it, find the set of possible values for e. [3]

(i) Sketch the Lines before you deal with the Simultaneous Equations

'Find the coordinates of the vertices of the triangle.'

You don't *have* to draw a sketch, but sketches always help make questions like this clearer.

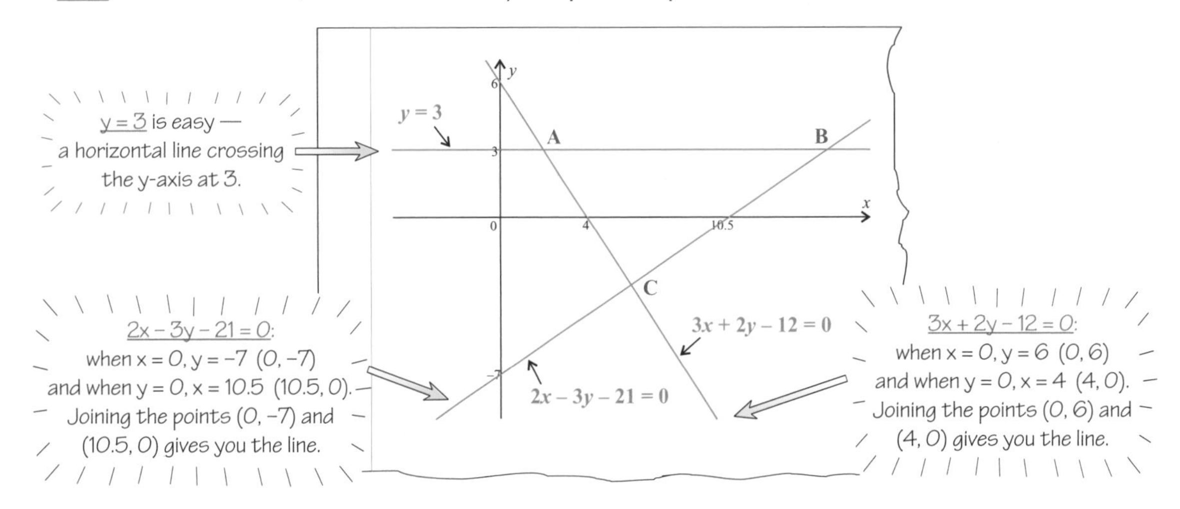

From the graph, you can see roughly where points A, B and C are, but you'll need to go through the algebra.

Finding A and B is straightforward, since you know that the y-coordinate is 3:

So for A:

$3x + 2y - 12 = 0$
$3x + (2 \times 3) - 12 = 0$
$3x + 6 - 12 = 0$
$3x = 6$
$x = 2$ **A has coordinates (2, 3).**

And for B:

$2x - 3y - 21 = 0$
$2x - 3 \times 3 - 21 = 0$
$2x - 9 - 21 = 0$
$2x = 30$
$x = 15$ **B has coordinates (15, 3).**

At C, you've got to solve a pair of simultaneous equations — bad luck. Label the equations: $2x - 3y - 21 = 0$ (1)
$3x + 2y - 12 = 0$ (2)

Multiply equation (1) by 2 and equation (2) by 3 to get the coefficients of y equal:

$4x - 6y - 42 = 0$
$9x + 6y - 36 = 0$

Now add together to get rid of y:

$13x - 78 = 0$
$13x = 78$
$x = 6$

Substitute into equation (1) or (2) to find y:

Using eq. (2):

$3x + 2y - 12 = 0$ (2)
$(3 \times 6) + 2y - 12 = 0$
$18 + 2y - 12 = 0$
$2y + 6 = 0$
$2y = -6$
$y = -3$ **C has coordinates (6, –3).**

These answers seem to make sense if you look back at the sketch, which is reassuring.

You should really check that works, by substituting your values into equation (1), but I'm running out of room.

Paper 2 Q3 — Simultaneous Equations

(ii) A Right Angle probably means you need the Gradient Rule

'Show that the triangle is right-angled.'

Whenever you see 'right angle' in a coordinate geometry question, remember that the gradients of two perpendicular lines multiply together to make –1. So if you can show that they do, you've shown the triangle must be right-angled.

To find the gradient of each line, rearrange it into the form $y = mx + c$ — and m will be the gradient.
You can tell from the diagram that the right angle looks like it's at C, so use BC and AC.

Line BC: $2x - 3y - 21 = 0$
$3y = 2x - 21$
divide by 3: $y = \frac{2}{3}x - 7$

Line AC: $3x + 2y - 12 = 0$
$2y = -3x + 12$
divide by 2: $y = -\frac{3}{2}x + 6$

You can see from the $y = mx + c$ form of the equations that the gradients are $\frac{2}{3}$ and $-\frac{3}{2}$.

$\frac{2}{3} \times -\frac{3}{2} = -1$ so the lines that meet at C are perpendicular, and the triangle must be right-angled.

(iii) You know what I'm gonna say — sketch it on your graph

'If point D lies outside the triangle, but not on it, show that either $d > 3$ or $d < 1.5$.'

Point D has co-ordinates (3, d), so it must lie on the line x = 3. It lies outside the triangle, so you can see from the sketch that it must lie **above the line AB** or **below the line AC**.

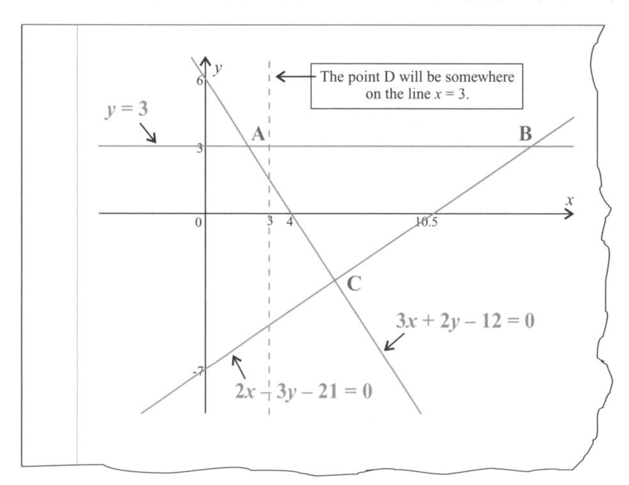

First off you can happily say that $d > 3$ is one condition, because $y = 3$ is the top of the triangle.

Secondly you have to work out the values of d that would give D is below the line AC.

If D was on the line AC, then $(3, d)$ would satisfy $3x + 2y - 12 = 0$:

$3 \times 3 + 2d - 12 = 0$
$9 + 2d - 12 = 0$
$2d - 3 = 0$
$d = 1.5$

But as D can't be on the triangle, you need the y-coordinate to be less than 1.5, i.e. $d < 1.5$.

So the conditions are $d > 3$ or $d < 1.5$.

(iv) Same thing again... almost

'Given that E lies inside the triangle, but not on it, find the set of possible values for e.'

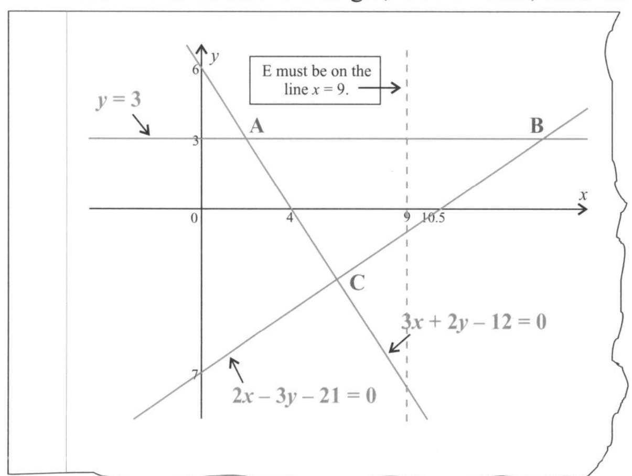

From the diagram you can tell that E is below AB and above BC. So, clearly, $e < 3$. Now work out the values of e so that E is above the line BC.

If e was on the line BC, then $(9, e)$ would satisfy $2x - 3y - 21 = 0$:

$2x - 3y - 21 = 0$
$2 \times 9 - 3e - 21 = 0$
$18 - 3e - 21 = 0$
$-3e - 3 = 0$
$e = -1$

So above the line $e > -1$

Putting both conditions together gives you: $-1 < e < 3$

Simultaneous equations and graphs and gradients and inequalities...

These are a few of my favourite things.

Paper 2 Q4 — Integration and Graphs

> **4** A curve has equation $y = \mathrm{f}(x)$, where $dy/dx = 4(1 - x)$.
> The curve passes through the point A, with coordinates (2, 6).
>
> **(i)** Find the equation of the curve. [4]
>
> **(ii)** Sketch the graph of $y = \mathrm{f}(x)$. [3]

(i) *Time to Integrate*

'Find the equation of the curve.'

Well let me see... you're given dy/dx but not y...
I'd wager that you have to integrate that one, my friend.

But you've got a bracket in the way, so multiply that out first, to get each term as a power of x:

Now integrate each side:

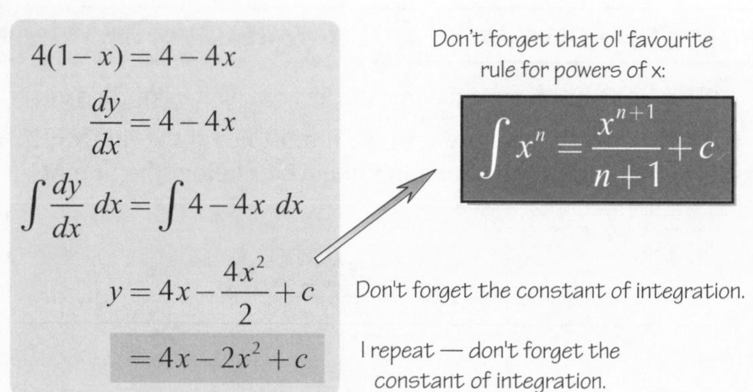

$$4(1-x) = 4-4x$$

$$\frac{dy}{dx} = 4-4x$$

$$\int \frac{dy}{dx}\,dx = \int 4-4x\,dx$$

$$y = 4x - \frac{4x^2}{2} + c$$

$$= 4x - 2x^2 + c$$

$$\int x^n = \frac{x^{n+1}}{n+1} + c$$

Now you can work out what the value of c is using the known coordinate, (2, 6):

You can substitute these values of x and y into the equation you've got so far.

At A: $x = 2$ and $y = 6$

$y = 4x - 2x^2 + c$

$6 = 4 \times 2 - 2 \times 2^2 + c$

$6 = 8 - 8 + c$

$c = 6$

So that gives: $y = 4x - 2x^2 + 6$

(ii) *OK class — crayons out...*

'Sketch the graph of $y = \mathrm{f}(x)$.'

It's a quadratic, like you've seen a million times before — so you should know the shape.
But you still need to know:

1) the stationary point (where the gradient = 0)
2) whether it's a maximum or minimum
3) any points where it cuts the axes

Paper 2 Q4 — Integration and Graphs

1) You've already got dy/dx, so it's a quick job working out the stationary point.

$$\frac{dy}{dx} = 4 - 4x$$

At a stationary point: $\frac{dy}{dx} = 0 \Rightarrow 4 - 4x = 0$

$4 = 4x$

$x = 1$

And substituting $x = 1$ into the equation for y gives:

$y = 4x - 2x^2 + 6$

$= 4 \times 1 - 2 \times 1^2 + 6$

$= 4 - 2 + 6 = 8$

So the stationary point is at: $(1, 8)$

2) Maximum or minimum? — Look at the sign of the x^2 term.

You've got $y = 4x - 2x^2 + 6$, which has negative x^2 term — so you know that the parabola will be n-shaped instead of u-shaped. In other words, **(1, 8)** is a **maximum**.

3) All that's left is to find the points where it cuts the axes.

Where it cuts the y-axis, $x = 0$.

That gives: $y = 4x - 2x^2 + 6$

$y = 4 \times 0 - 2 \times 0^2 + 6$

$y = 6$

So it crosses the y-axis at: $(0, 6)$

Where it cuts the x-axis, $y = 0$.

That gives: $y = 4x - 2x^2 + 6$

$0 = 4x - 2x^2 + 6$

Rearrange to give a positive x^2: $2x^2 - 4x - 6 = 0$

This can be factorised: $2(x^2 - 2x - 3) = 0$

$2(x + 1)(x - 3) = 0$

So the solutions will be at: $x = -1$ and $x = 3$

...so the curve crosses the x-axis at: $(-1, 0)$ and $(3, 0)$

And that's enough information for you to do the sketch:

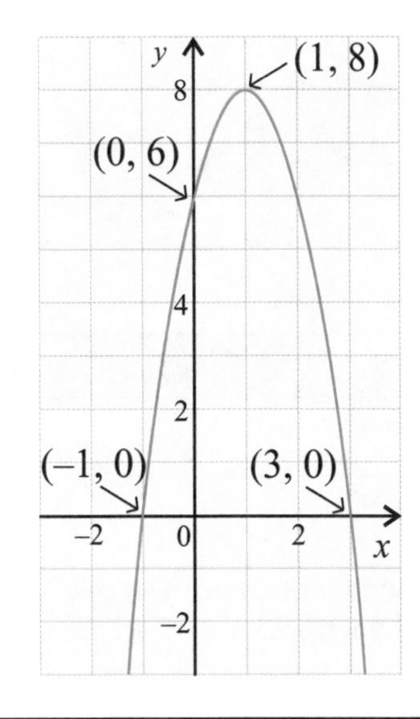

All the key-points clearly labelled...
...beautiful.

I'm waiting for parabola hats to come into fashion...

The constant of integration — a boring little '+ c', just waiting to be forgotten. If you find your equation isn't going anywhere, check you've tacked it on the end. It mightn't seem terribly important, but the examiners reckon it's a big deal, and it's generally best to pander to their every whim.

Paper 2 Q5 — Algebraic Fractions

5 **(i)** Solve the equation: $\frac{4}{(x-2)}=\frac{6}{(2x+5)}$. [3]

(ii) Show that: $\frac{4}{9(x+2)}+\frac{1}{3(x-1)^2}+\frac{5}{9(x-1)}=\frac{x^2}{(x+2)(x-1)^2}$. [5]

(i) Get rid of Denominators when you can

'Solve the equation: $\frac{4}{(x-2)}=\frac{6}{(2x+5)}$.'

Whenever you have equations with fractions, it's a good idea to multiply each side of the equation by the denominators.

Here's what you've got to solve: $\frac{4}{(x-2)}=\frac{6}{(2x+5)}$

You've got **2** denominators, so multiply each side by $(x-2)$ **and** $(2x+5)$.

$$\frac{4(2x+5)(x-2)}{(x-2)}=\frac{6(2x+5)(x-2)}{(2x+5)}$$

$$4(2x+5)=6(x-2)$$

This is cross-multiplication. Multiplying by the denominators of both sides will always get rid of the fractions.

Now multiply out and solve the equation:

$$4(2x+5)=6(x-2)$$
$$8x+20=6x-12$$
$$2x=-32$$
$$x=-16$$

Check by subbing in this value of x into original equation:

$$\frac{4}{(-16-2)}=\frac{6}{(2\times(-16)+5)}$$
$$\frac{4}{-18}=\frac{6}{-27}$$
$$-\frac{2}{9}=-\frac{2}{9}\ ✓$$

(ii) Adding fractions? — You need a Common Denominator

'Show that: $\frac{4}{9(x+2)}+\frac{1}{3(x-1)^2}+\frac{5}{9(x-1)}=\frac{x^2}{(x+2)(x-1)^2}$.'

This looks unlikely at first glance, but don't let that put you off. When you add algebraic fractions, the technique is the same as for normal fractions. You need a **common denominator**.

You basically need to multiply together **all** the denominators on the LHS.
But **when one term is a factor of another term**, you only need the **larger term**.

This only really makes sense when you see it work in practice...

Paper 2 Q5 — Algebraic Fractions

Finding the common denominator:

Look at the numbers in the LHS denominators. You've got 9, 3 and 9. Well, 3 is a factor of 9, so you only need the larger term, i.e. **9**.

The next thing to take a look at is the **$(x + 2)$**. This term **isn't** a factor of any of the other terms on the bottom, so you **will** need this.

Then you've got $(x - 1)^2$. One of the fractions has $(x - 1)$ as its denominator, which is of course a factor of $(x - 1)^2$. So again you just need the larger term, i.e. $\mathbf{(x - 1)^2}$.

You've now looked at all the terms in the three denominators, and the common denominator you're left with is: $9(x + 2)(x - 1)^2$

Make the denominator of each part the common denominator:

This involves multiplying the top and bottom of each fraction by **the terms of the common denominator that aren't in the denominator of the original fraction**. I know this sounds horrendous, but it'll make more sense when you see it in action.

Look at this fraction on the left: you've already got $9(x + 2)$ on the bottom, so you only need to multiply top and bottom by $(x - 1)^2$.

$$\frac{4}{9(x+2)}+\frac{1}{3(x-1)^2}+\frac{5}{9(x-1)}$$

$$=\frac{4(x-1)^2}{9(x+2)(x-1)^2}+\frac{3(x+2)}{9(x+2)(x-1)^2}+\frac{5(x+2)(x-1)}{9(x+2)(x-1)^2}$$

You should be able to cancel down all the new fractions back to the original fractions. E.g. $\frac{5(x+2)(x-1)}{9(x+2)(x-1)^2}=\frac{5}{9(x-1)}$

Now multiply out the top bits.

$$=\frac{4x^2-8x+4}{9(x+2)(x-1)^2}+\frac{3x+6}{9(x+2)(x-1)^2}+\frac{5x^2+5x-10}{9(x+2)(x-1)^2}$$

Now you can turn it into just one fraction, and then it's just a question of collecting all the like terms.

$$=\frac{4x^2-8x+4}{9(x+2)(x-1)^2}+\frac{3x+6}{9(x+2)(x-1)^2}+\frac{5x^2+5x-10}{9(x+2)(x-1)^2}$$

$$=\frac{4x^2+5x^2-8x+3x+5x+4+6-10}{9(x+2)(x-1)^2}$$

$$=\frac{9x^2}{9(x+2)(x-1)^2}$$

The 9s cancel down: $=\frac{x^2}{(x+2)(x-1)^2}$ ⟸ and this is what the question asks for.

'The Denominator' — starring Armando Schwarzkopf...

Algebraic fractions — I shudder at the very sight of them. The most awkward bit when you're adding them, though, is finding the lowest common denominator. If it comes to the day and you can't remember how to do it, just multiply all the denominators together and use that. It's less elegant, but it *is* a common denominator.

Paper 2 Q6 — Differentiation and Graphs

6 A curve has the equation $y = f(x)$, where $f(x) = x^3 - 3x + 2$.

(i) Find dy/dx. [2]

(ii) Find the points at which the gradient of the curve $y = f(x)$ is 0. [4]

(iii) Show that $x^3 - 3x + 2$ factorises to $(x - 1)^2(x + 2)$. [2]

(iv) Sketch the graphs of:

(a) $y = f(x)$ [3]

(b) $y = f(x - 3)$ [2]

(c) $y = 2f(x)$ [2]

(i) Differentiate each term as usual

'Find dy/dx.'

For powers of x you **multiply by the power**, then **drop the power by 1**.
And you do that for each term.

$$y = x^3 - 3x + 2$$

$$\frac{dy}{dx} = 3x^{3-1} - 3x^{1-1} + 0$$

$$= 3x^2 - 3$$

$x^0 = 1$

You should be able to differentiate in your sleep by now.

(ii) Put dy/dx = 0

'Find the points at which the gradient of the curve $y = f(x)$ is 0.'

So, you need to find where $dy/dx = 0$. You've just worked out dy/dx — so just set that equal to 0:

$$\frac{dy}{dx} = 3x^2 - 3 = 0$$

You can take a factor of 3 out, leaving: $x^2 - 1 = 0$

You've got 'x^2 – a number', so you can use the difference of two squares:

$$(x + 1)(x - 1) = 0$$

Putting each bracket equal to 0 gives the solutions:

$$x = -1 \text{ and } x = 1$$

Not finished yet though. The question asks for the points, so you've got to give coordinates.
That means you need the y-values as well:

When $x = -1$:
$y = x^3 - 3x + 2$
$= (-1)^3 - (3 \times -1) + 2$
$= -1 + 3 + 2$
$= 4$

When $x = 1$:
$y = x^3 - 3x + 2$
$= 1^3 - (3 \times 1) + 2$
$= 1 - 3 + 2$
$= 0$

So the points are at: $(-1, 4)$ and $(1, 0)$.

Paper 2 Q6 — Differentiation and Graphs

(iii) Don't make a meal of this...

'Show that $x^3 - 3x + 2$ factorises to $(x - 1)^2(x + 2)$.'

You've been given all the factors, so the easiest way to check they're right is to multiply them out.

Start by expanding $(x - 1)^2$.

When you square a bracket, it's a good idea to write it out in full, otherwise you might forget about the terms in x.

$$(x-1)^2 = (x-1)(x-1) = x^2 - x - x + 1 = x^2 - 2x + 1$$

now multiply by $(x + 2)$: $(x-1)^2(x+2) = (x^2 - 2x + 1)(x + 2)$

multiply out RHS: $= x^3 + 2x^2 - 2x^2 - 4x + x + 2$

$= x^3 - 3x + 2$

Stop and check — if you multiply a bracket with 3 things in by a bracket with 2 things in, you should get 6 things altogether.

And that's the right answer.

(iv)(a) Sketching — at last something I'm good at...

'Sketch the graph of $y = f(x)$.'

This bit uses all of the previous bits — as is often the way with these huge, multi-part questions.

To sketch a curve, you'd ideally like to know:

1) any turning points
2) where it crosses the axes
3) a rough idea of the shape

1) You know that the turning points are at (–1, 4) and (1, 0) from part (ii).

2) You're already on your way to finding the x-intercepts, from the factorising you did for part (iii).

When $y = 0$: $(x-1)^2(x+2) = 0$, so $x = 1$ and $x = -2$

So the x-intercepts are (1, 0) and (-2, 0).

When $x = 0$: $y = x^3 - 3x + 2 = 0 - 0 + 2 = 2$

So the y-intercept is (0, 2).

3) You should have a rough idea of the shape of this cubic. It has a positive x^3 term so it will start low and finish high, and you know where the two turning points are.

Putting all these together, you can now sketch the graph.

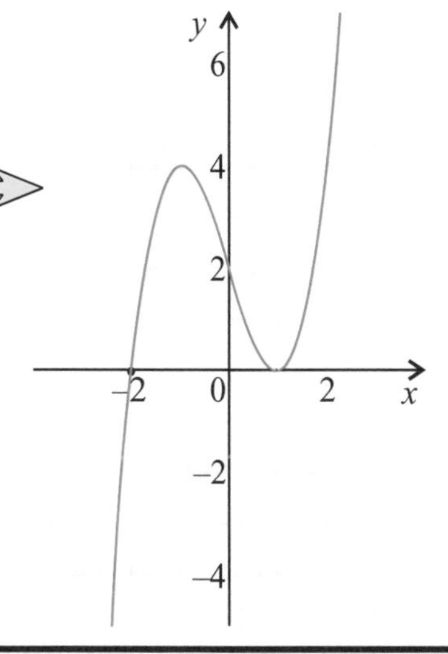

Paper 2 Q6 — Differentiation and Graphs

(iv)(b) Translate the graph

'Sketch the graph of $y = f(x - 3)$'

There are a few types of transformation you've just got to **learn**.
This one's pretty easy — just a simple **translation**.
The graph of $\mathbf{f(x - a)}$ is the same as the graph of $f(x)$, but **shifted to the right**.

To get $f(x - 3)$, the x-value of every point needs to be increased by 3.

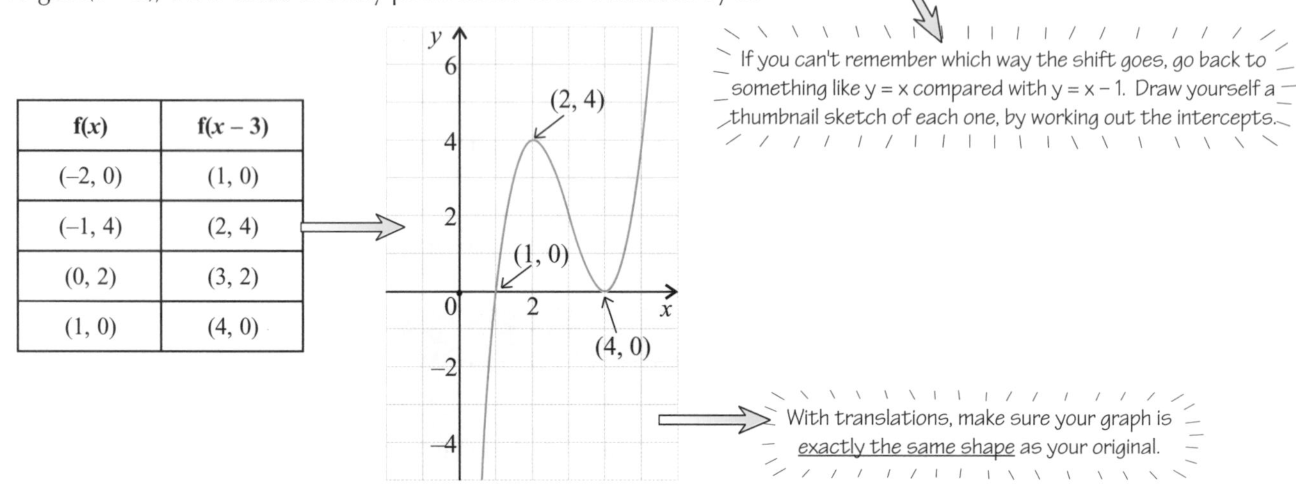

f(x)	f(x – 3)
(–2, 0)	(1, 0)
(–1, 4)	(2, 4)
(0, 2)	(3, 2)
(1, 0)	(4, 0)

(iv)(c) And 1 and 2 and... Stretch...

'Sketch the graph of $y = 2f(x)$'

This time it's a **stretch** up the ***y*-axis**, with **scale factor 2**.
This is pretty straightforward — all you need to do is **double the *y*-coordinates** of the original graph.

You could do a detailed table of values, but it's enough just to look at the points you worked out in part (iv)(a).

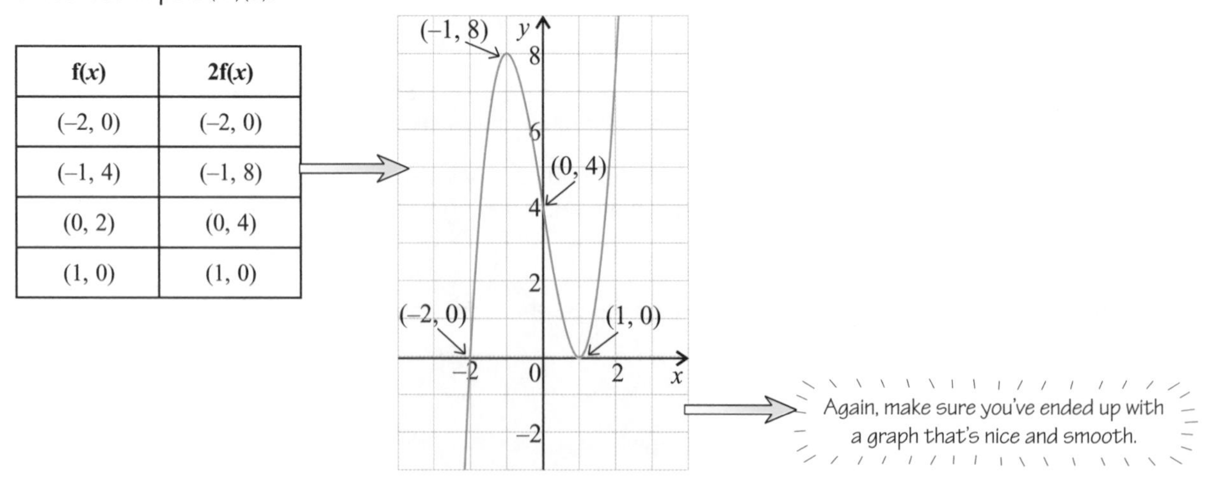

f(x)	2f(x)
(–2, 0)	(–2, 0)
(–1, 4)	(–1, 8)
(0, 2)	(0, 4)
(1, 0)	(1, 0)

Translate 'a graph' — 'una gráfica', 'un graphique', 'ein Schaubild'...

This might seem a bit 'stating the obvious'-y, but you need to read the question really carefully. If they ask you for a "point", they want both the x- and y-coordinates. In part ii), if you'd only given the x-coordinates, you'd have only got half the marks.

Paper 2 Q7 — Cubic Equations

7 A function is defined by $f(x) = x^3 - 4x^2 - 7x + 10$.

$(x - 1)$ is a factor of $f(x)$.
Hence or otherwise solve the equation $x^3 - 4x^2 - 7x + 10 = 0$. [2]

To solve the equation you must factorise completely

'$(x - 1)$ is a factor of $f(x)$.
Hence or otherwise solve the equation $x^3 - 4x^2 - 7x + 10 = 0$'

Whenever you see "hence", you should be making use of part of the question.

You know that one factor of $x^3 - 4x^2 - 7x + 10$ is $(x - 1)$, so now you have to try and find some more.

The cubic expressions you meet can always be factorised into a **linear** and **quadratic** factor. You already know a linear factor, so the factorising is going to look something like this:

$$(x - 1)(\ldots\ldots\ldots\ldots\ldots\ldots) = x^3 - 4x^2 - 7x + 10$$

The empty bracket will be a quadratic expression. The first and last terms are easy to find:

$$(x - 1)(1x^2 \;\ldots\ldots\; ?x \;\ldots\ldots\; - 10) = x^3 - 4x^2 - 7x + 10$$

these multiply to give... these multiply to give...

Now you need to get $-4x^2$:

-1 times $1x^2$ will give $-x^2$, so you need another $-3x^2$.
x times $?x$ makes $-3x^2$ so $? = -3$

You can check that this will give the right number of x: $x \times -10 = -10x$
and: $-1 \times -3x = +3x$
giving: $-7x$ altogether

So far then we have: $(x - 1)(x^2 - 3x - 10) = 0$

The quadratic part will factorise: $(x - 1)(x - 5)(x + 2) = 0$

Put each bracket equal to 0 to find the solutions: $x = 1, 5$ or -2

2001 = x^3 – 4: Stanley Cubic's maths in space masterpiece...

Factorising a cubic is tricky, but you can always check your answer by multiplying out again. And that's a **really good idea**.

Paper 2 Q8 — Indices

8 **(i)** What is the exact value of 5^{-2}? [1]

(ii) What is the exact value of $(4^{\frac{1}{2}})^6$? [1]

(iii) Simplify the expression: $\dfrac{x^3 \times x^4}{\sqrt{x^{10}}}$ [2]

(i) Rearrange negative powers

'What is the exact value of 5^{-2}?'

When you see a negative power it is usually best to rewrite it as a positive power.

The important rule for negative powers is:

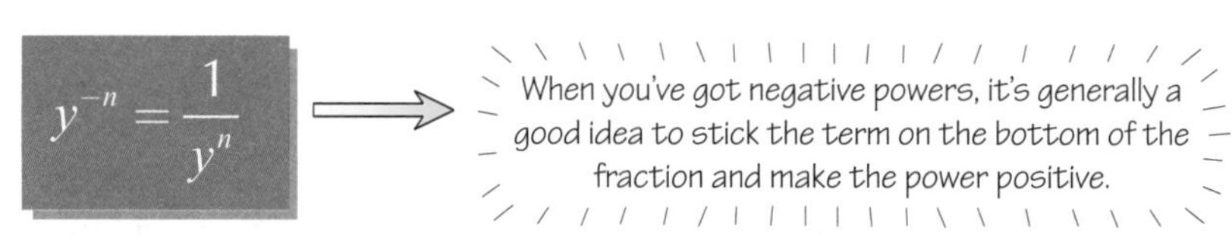

Applying this rule:

$$5^{-2} = \frac{1}{5^2} = \frac{1}{25}$$

(ii) Remember your Power Laws

'What is the exact value of $(4^{\frac{1}{2}})^6$?'

The rule for something raised to a power then raised to another power is:

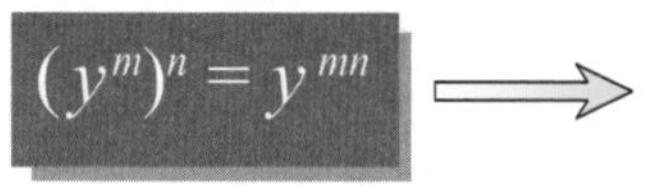

When you've got a bracket that's raised to a power, you can get rid of the bracket by multiplying the power outside the bracket by the power(s) inside the bracket.

Once again, just apply the rule:

$$(4^{\frac{1}{2}})^6 = 4^{\frac{1}{2}\times 6}$$

$$(4^{\frac{1}{2}})^6 = 4^3$$

$$(4^{\frac{1}{2}})^6 = 64$$

By the way, you can still work this out if you don't know the rule — just work out the bracket first:

$$4^{\frac{1}{2}} = \sqrt{4} = 2$$

So: $(4^{\frac{1}{2}})^6 = 2^6$

and then: $2^6 = 2 \times 2 \times 2 \times 2 \times 2 \times 2 = 64$

Paper 2 Q8 — Indices

(iii) Work out the top and bottom — then Divide

'Simplify the expression: $\dfrac{x^3 \times x^4}{\sqrt{x^{10}}}$,

As you would with any complicated fraction, simplify the top first.
Another power law to remember:

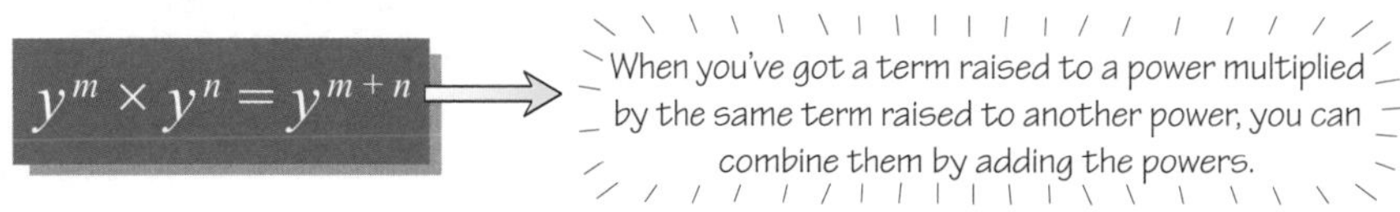

So we have: $x^3 \times x^4 = x^{3+4} = x^7$

That's the top part sorted. Now for the bottom part. Remember the square root is the same as the power ½.

$$\sqrt{x^{10}} = (x^{10})^{\frac{1}{2}}$$

This is the same type of expression as in part (ii).

Using: $(y^m)^n = y^{mn}$

$$(x^{10})^{\frac{1}{2}} = x^{10 \times \frac{1}{2}} = x^5$$

So you've got the top and bottom of the fraction. You've simplified the expression to: $\dfrac{x^7}{x^5}$.

Now you need to remember what happens when you divide powers...

$$y^m \div y^n = y^{m-n}$$

When you've got a term raised to a power divided by the same term raised to another power, you can combine them by subtracting the powers.

So: $\dfrac{x^7}{x^5} = x^{7-5} = x^2$

Who'd have thought that something that looked so messy could end up being something so simple.

Indices — an endless source of joy... ☺

There's no excuse for getting this stuff wrong. Learn the laws, get heaps of practice and you'll be fine on the day.
There are no clever tricks, you just follow the rules and the answer will get churned out the other end.

Paper 2 Q9 — Tangents and Normals

9 The diagram shows part of the graph with equation $y = x^3 - 2x^2 + 4$.

(i) Find the equation of the tangent at $x = 1$. [3]

(ii) Find the equation of the normal to the curve at $x = 2$. [3]

(iii) Find the distance between where the tangent and the normal cross the x-axis. [2]

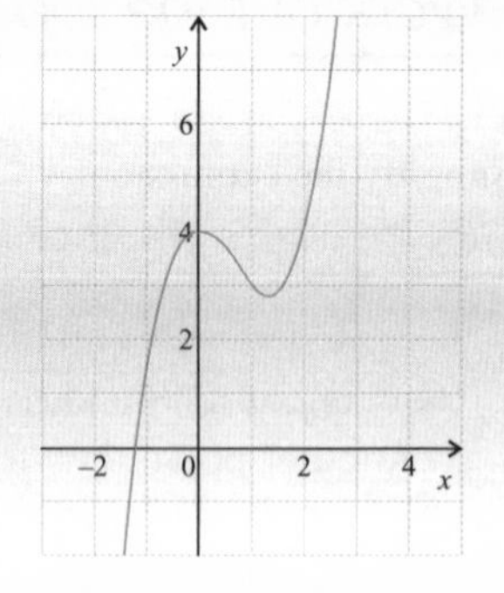

(i) First find dy/dx

'Find the equation of the tangent at $x = 1$.'

They've given you a sketch of the line, so you can just sketch the tangent straight onto it. The tangent at $x = 1$ is a straight line with the same gradient as the curve at that point.

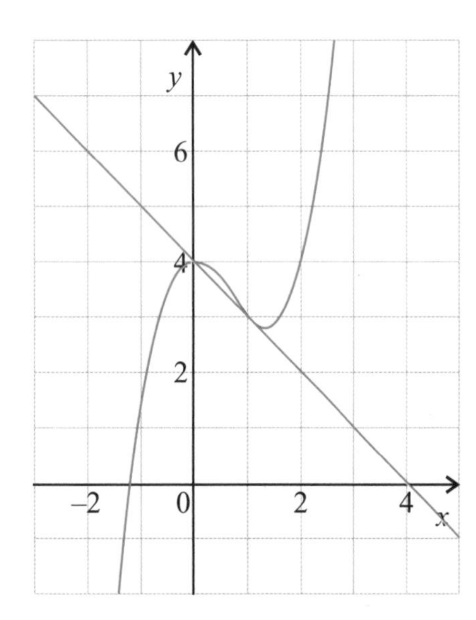

To get the gradient of the curve and tangent you need to differentiate — that means finding dy/dx.

Use the rule for powers of x: $\frac{d}{dx}(x^n) = nx^{n-1}$

$$y = x^3 - 2x^2 + 4$$

$$\frac{dy}{dx} = 3x^2 - 4x$$

Now put in the value of x.

At $x = 1$: $\frac{dy}{dx} = 3 \times 1^2 - 4 \times 1$

$$\frac{dy}{dx} = 3 - 4 = -1$$

So the gradient at $x = 1$ is -1.

To find the equation of the tangent you can use: $y - y_1 = m(x - x_1)$

You know the gradient, and it's easy to find a point on the line by putting a simple value of x back into the equation (when $x = 1$, $y = 3$).

$$y - y_1 = m(x - x_1)$$

$$y - 3 = -1(x - 1)$$

$$y - 3 = -x + 1$$

$$y = -x + 4$$

'$y + x = 4$' is also fine, of course.

Paper 2 Q9 — Tangents and Normals

(ii) I love Maths — does that make me normal?

'Find the equation of the normal to the curve at $x = 2$.'

You can add another doodle to the diagram.

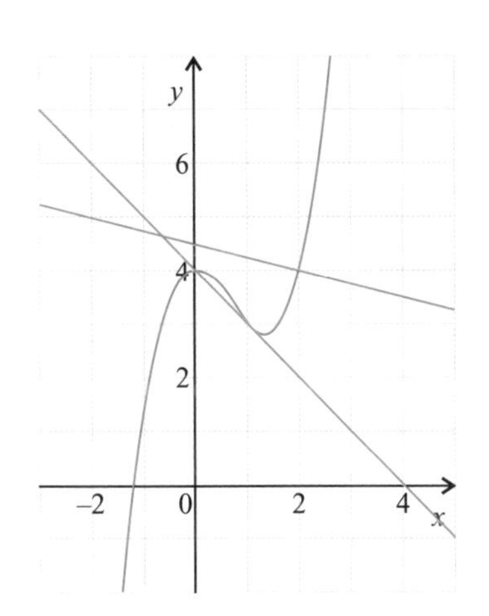

The normal at $x = 2$ is the line at right angles to the tangent at that point. You can therefore use the gradient rule:

$$\text{Gradient} = \frac{-1}{\text{gradient of a perpendicular line}}$$

So you can use the differential you found in part (i) to get the gradient of the tangent, and then use this rule to find the normal. Sounds like a plan to me.

$$\frac{dy}{dx} = 3x^2 - 4x$$

At $x = 2$:

$$\frac{dy}{dx} = 3 \times 2^2 - 4 \times 2$$

$$\frac{dy}{dx} = 12 - 8 = 4$$

This is the gradient of the tangent.

Therefore, using the gradient rule: Gradient of normal $= \frac{-1}{4}$

You know the gradient and a point on the line (by putting $x = 2$ into the equation you get $y = 4$), so you can use my favourite rule again:

$$y - y_1 = m(x - x_1)$$

$$y - 4 = -\frac{1}{4}(x - 2)$$

$$y - 4 = -\frac{x}{4} + \frac{1}{2}$$

$$y = -\frac{x}{4} + \frac{9}{2}$$

(iii) Find the intercepts on the x-axis

'Find the distance between where the tangent and the normal cross the x-axis.'

To get the x-axis intercepts, you need to put $y = 0$ into each equation.

$$y = -x + 4$$

$$0 = -x + 4$$

$$x = 4$$

$$y = -\frac{x}{4} + \frac{9}{2}$$

$$0 = -\frac{x}{4} + \frac{9}{2}$$

$$\frac{x}{4} = \frac{9}{2}$$

$$x = 18$$

The lines therefore cross the x-axis at: (4, 0) and (18, 0)

So the distance between the two intercepts is:

$18 - 4 = 14$.

I wouldn't normally do this kind of thing...

Aaaah... it's another one of those 'tangent-has-the-same-gradient-as-the-curve, find-the-gradient-of-the-curve, find-the-equation-of-the-tangent, use-the-tangent/normal-gradient-rule, find-the-gradient-of-the-normal, find-the-equation-of-the-normal, find-the-intercepts, do-a-subtraction' questions. Don't you just love 'em.

Paper 2 Q10 — Arithmetic Series

10 (i) A sequence is defined by the recurrence relation $x_{n+1} = 3x_n - 4$.
Given that the first term is 6, find x_4. [2]

(ii) The third term of an arithmetic progression is 9 and the seventh term is 33.
(a) Find the first term and the common difference. [3]
(b) Find S_{12}, the sum of the first 12 terms in the series. [3]
(c) Hence or otherwise find: $\sum_{1}^{12}(6n+1)$ [2]

(i) Work out the Terms of the Sequence one at a time

'Given that the first term is 6, find x_4.'

The words 'recurrence relation' make it sound complicated, but it's just a fancy way of saying 'this is how you get the next term of the sequence'.

The rule for this sequence is $x_{n+1} = 3x_n - 4$.

That means to get the next term, you multiply the one you've got by 3, then subtract 4.

You know the first term, so you have to just work them out one by one:

$$x_1 = 6$$
$$x_2 = 3x_1 - 4 = 3 \times 6 - 4 = 14$$
$$x_3 = 3x_2 - 4 = 3 \times 14 - 4 = 38$$
$$x_4 = 3x_3 - 4 = 3 \times 38 - 4 = 110$$

And you can stop there — you only need x_4. $x_4 = 110$ is your answer.

(ii)(a) Use the General Expression for the nth Term in an AP

'Find the first term and the common difference.'

The general expression for the nth term in an arithmetic progression is: $u_n = a + (n - 1)d$

You can use this to write some equations:

3rd term = 9
$u_3 = a + 2d = 9$

7th term = 33
$u_7 = a + 6d = 33$

So you've got:
$a + 2d = 9$
$a + 6d = 33$

...which looks suspiciously like simultaneous equations — ahahahahaha.

But happily they're pretty easy.

Subtracting the top one from the bottom one gives:

Substituting into either one of the equations gives:

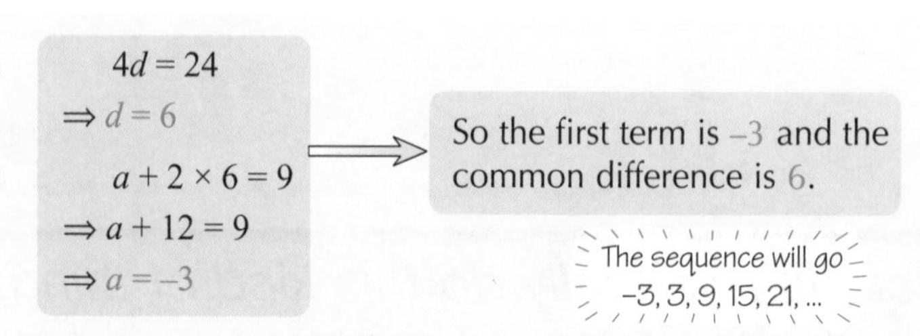

Paper 2 Q10 — Arithmetic Series

(ii)(b) Use the formula for the sum of a series

'Find S_{12}, the sum of the first 12 terms in the series.'

This is just a case of sticking the numbers into the sum formula: $S_n = \frac{n}{2}[2a + (n-1)d]$

For this series, you've got $a = -3$, $d = 6$ and $n = 12$.

$$S_{12} = \frac{12}{2}[2 \times -3 + (12-1) \times 6]$$
$$S_{12} = 6(-6 + 11 \times 6)$$
$$S_{12} = 6 \times 60$$
$$S_{12} = 360$$

Another formula you could use is $S_n = \frac{n}{2}(a+l)$. This one's best if you know l, the last term being added — but it's not so appropriate here. You could work it out for this series, but it'd take a lot longer. Best plan is to learn both formulas — then you can use the one that's most appropriate.

(ii)(c) Try to work out how it relates to the rest of the question

'Hence or otherwise find: $\sum_{1}^{12}(6n+1)$'

Don't be put off by the Greek letter sigma — it just means find the total.

Start by writing out some of the sequence, and see if anything springs to mind.

$6n + 1$ is the rule for finding the nth term of the sequence — so that gives you:

First term: $n = 1 \rightarrow 6n + 1 = 7$
Second term: $n = 2 \rightarrow 6n + 1 = 13$
Third term: $n = 3 \rightarrow 6n + 1 = 19$
...and so on.

It's actually another arithmetic progression with common difference of 6. Yay.

Except... the first term is 7 instead of –3. D'oh.
But all is not lost. All it means is that each term is 10 bigger than the corresponding term in the series from part (ii)(a):

Part (ii)(a) series:	–3	3	9	15	21
This series:	7	13	19	25	31

If each term is 10 bigger, the total of 12 terms will be 120 bigger than the sum of the series from part (ii)(a).

from part (ii)(b)

So: $\sum_{1}^{12}(6n+1) = 360 + 120 = 480$

It's a series business...

If you're finding the formulas difficult to remember, go back to Section 5 and see where they came from in the first place. If you understand why they are what they are, you could just work them out in the exam instead of having to rely on your memory. Some people find it easier to learn things like this parrot-fashion, some people don't. Up to you.

Answers

Section One — Algebra Fundamentals

1) a) a & b are constants, x is a variable.
b) k is a constant, θ is a variable.
c) a, b & c are constants, y is a variable.
d) a is a constant, x, y are variables

2) Identity symbol is $\equiv$.

3) A, C and D are identities.

4) a) -1.1 **b)** f is undefined (or ∞) **c)** 9.0 **d)** 0

5) a) x^8 **b)** a^{15} **c)** x^6 **d)** a^8 **e)** x^4y^3z **f)** $\frac{b^2c^5}{a}$

6) a) 4 **b)** 2 **c)** 8 **d)** 1 **e)** 1/7

7) a) $x=\pm\sqrt{5}$ **b)** $x=-2\pm\sqrt{3}$

8) a) $2\sqrt{7}$ **b)** $\frac{\sqrt{5}}{6}$ **c)** $3\sqrt{2}$ **d)** $\frac{3}{4}$

9) a) $\frac{8}{\sqrt{2}}=\frac{8}{\sqrt{2}}\times\frac{\sqrt{2}}{\sqrt{2}}=\frac{8\sqrt{2}}{2}=4\sqrt{2}$ **b)** $\frac{\sqrt{2}}{2}=\frac{\sqrt{2}}{(\sqrt{2})^2}=\frac{1}{\sqrt{2}}$
(there are other possible ways to do these questions)

10) $136+24\sqrt{21}$

11) $3-\sqrt{7}$

12) a) a^2-b^2 **b)** $a^2+2ab+b^2$
c) $25y^2+210xy$ **d)** $3x^2+10xy+3y^2+13x+23y+14$

13) a) $xy(2x+a+2y\sin x)$ **b)** $\sin^2 x(1+\cos^2 x)$
c) $8(2y+xy+7x)$ **d)** $(x-2)(x-3)$

14) a) $\frac{52x+5y}{60}$ **b)** $\frac{5x-2y}{x^2y^2}$ **c)** $\frac{x^3+x^2-y^2+xy^2}{x(x^2-y^2)}$

15) a) $\frac{3a}{2b}$ **b)** $\frac{2(p^2+q^2)}{p^2-q^2}$ **c)** 

Section Two — Quadratics & Polynomials

1) a) $(x+1)^2$ **b)** $(x-10)(x-3)$
c) $(x+2)(x-2)$ **d)** $(3-x)(x+1)$
e) $(2x+1)(x-4)$ **f)** $(5x-3)(x+2)$

2) a) $(x-2)(x-1)=0$, so $x=1$ or 2
b) $(x+4)(x-3)=0$, so $x=-4$ or 3
c) $(2-x)(x+1)=0$, so $x=2$ or -1
d) $(x+4)(x-4)=0$, so $x=4, -4$
e) $(3x+2)(x-7)=0$, so $x=-2/3$ or 7
f) $(2x+1)(2x-1)=0$, so $x=\pm1/2$
g) $(2x-3)(x-1)=0$, so $x=1$ or $3/2$

3) a) $(x-2)^2-7$; minimum value $=-7$ at $x=2$, and this crosses the x-axis at $x=2\pm\sqrt{7}$
b) $\frac{21}{4}-\left(x+\frac{3}{2}\right)^2$; maximum value $=21/4$ at $x=-3/2$, and this crosses the x-axis at $x=-\frac{3}{2}\pm\sqrt{\frac{21}{4}}$
c) $2(x-1)^2+9$; minimum value $=9$ at $x=1$, and this doesn't cross the x-axis.
d) $4\left(x-\frac{7}{2}\right)^2-1$; minimum value $=-1$ at $x=7/2$, crosses the x-axis at $x=\frac{7}{2}\pm\frac{1}{2}$ i.e. $x=4$ or 3

4) a) $b^2-4ac=16$, so 2 roots
b) $b^2-4ac=0$, so 1 root
c) $b^2-4ac=-8$, so no roots

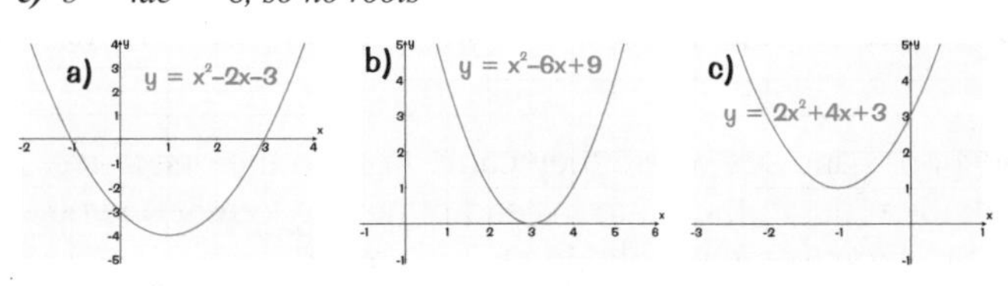

5) a) $x=1.77, 0.57$ **b)** $x=3.30, -0.30$ **c)** $x=1.16, -5.16$

6) $k^2-(4\times1\times4)>0$, so $k^2>16$ and so $k>4$ or $k<-4$.

7) First, complete the square to give:

$$a\left(x+\frac{b}{2a}\right)^2-\frac{b^2}{4a}+c=0$$

Then, rearrange to get x on one side:

$$\left(x+\frac{b}{2a}\right)^2=\frac{b^2}{4a^2}-\frac{c}{a}=\frac{b^2-4ac}{4a^2}$$

$$x=-\frac{b}{2a}\pm\frac{\sqrt{b^2-4ac}}{2a}=\frac{-b\pm\sqrt{b^2-4ac}}{2a}$$

Section Three — Simultaneous Equations and Inequalities

1) a) $x>-\frac{38}{5}$ **b)** $y\le\frac{7}{8}$ **c)** $y\le-\frac{3}{4}$

2) (i) $x>5/2$ **(ii)** $x>-4$ **(iii)** $x\le-3$

3) a) $-\frac{1}{3}\le x\le2$ **b)** $x<1-\sqrt{3}$ or $x>1+\sqrt{3}$ **c)** $x\le-3$ or $x\ge-2$

4) (i) $x\le-3$ and $x\ge1$ **(ii)** $x<-½$ and $x>1$
(iii) $-3<x<2$

5) a) $(-3,-4)$ **b)** $\left(-\frac{1}{6},-\frac{5}{12}\right)$

6) a) The line and the curve meet at the points $(2,-6)$ and $(7,4)$.
b) The line is a tangent to the parabola at the point $(2,26)$.
c) The equations have no solution and so the line and the curve never meet.

7) a) $\left(\frac{1}{4},-\frac{13}{4}\right)$ **b)** $(4,5)$ **c)** $(-5,-2)$

Section Four — Coordinate Geometry and Graphs

1) a) (i) $y+1=3(x-2)$ **(ii)** $y=3x-7$ **(iii)** $3x-y-7=0$
b) (i) $y+\frac{1}{3}=\frac{1}{5}x$ **(ii)** $y=\frac{1}{5}x-\frac{1}{3}$ **(iii)** $3x-15y-5=0$

2) a) $y=\frac{3}{2}x-4$ **b)** $y=-\frac{1}{2}x+4$

3) The equation of the required line is $y=\frac{3}{2}x+7.5$.

4)

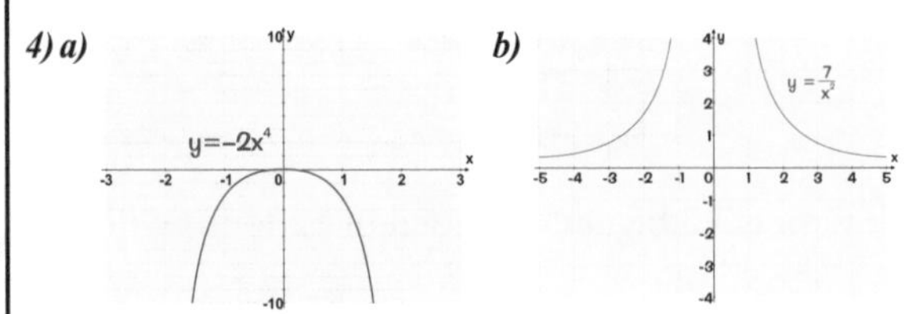

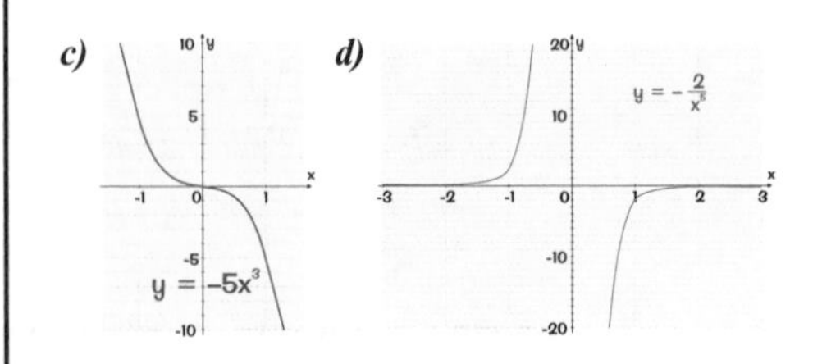

Answers

5)

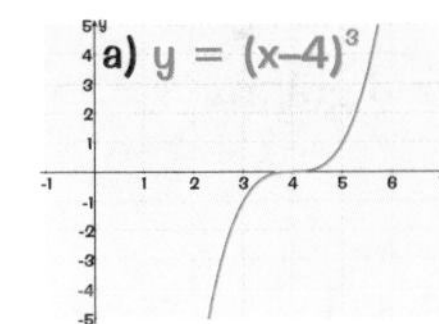

b) y = (3–x)(x+2)²

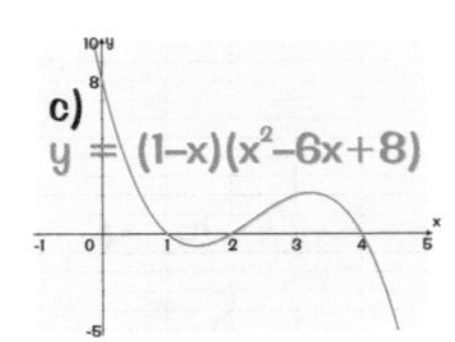

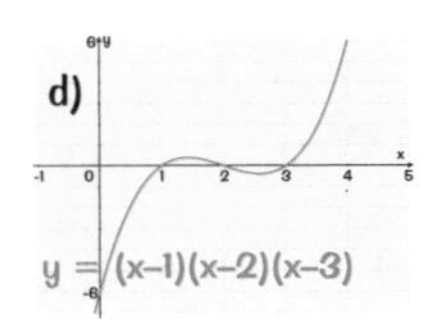

6) a)

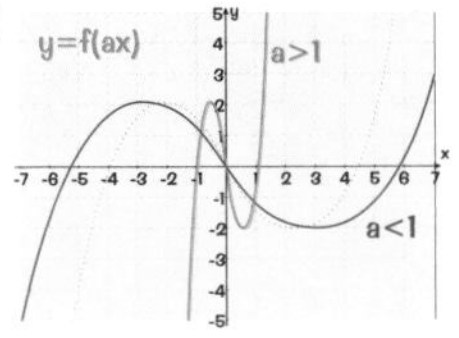

b)

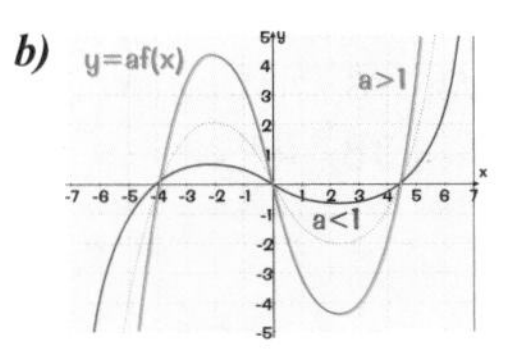

c)

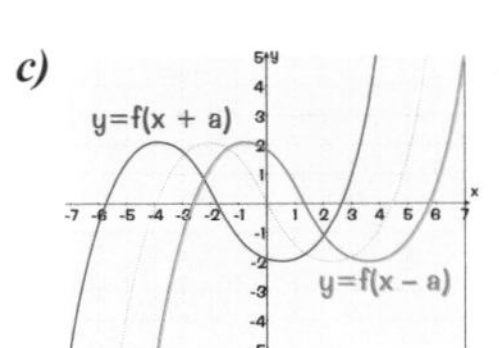

d)

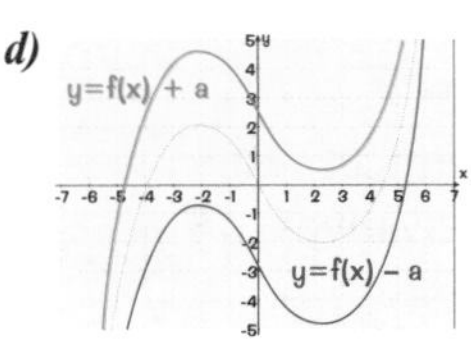

Section Five — Sequences and Series

1) a) *nth term = 4n – 2* **b)** *nth term = 0.5n – 0.3*
c) *nth term = –3n + 24* **d)** *nth term = –6n +82*

2) $a_{k+1} = a_k + 5,\ a_1 = 32$

3) *Last term (l) = 15*

4) *Common differrence (d) = 0.75*

5) $S_8 = 168$

6) *First work out n:* $a = 5,\ l = 65,\ d = 3$
so, $65 = 5 + 3(n - 1)$
so, $n = 21$
Now use: $S_{21} = 21 \times \frac{(5+65)}{2}$
so, $S_{21} = 735$

7) a) *Common difference (d) = 4*
b) *15th term = 63* **c)** $S_{10} = 250$

8) *First, work out d:*
We know that the 7th term is 36 and the 10th is 30.
So the differrence in 3 'moves' is –6 i.e. d = –2
This gives the expression for the nth term as –2n + 50
So, a = 48 (i.e. when n=1)
and, l = 40 (n = 5)
Now find the sum of the series: $S_5 = 5 \times \frac{(48+40)}{2}$ *so,* $S_5 = 220$

9) $S_{20} = 610$

10) $S_{10} = 205$

Section Six — Differentiation

1) $\frac{d}{dx}\left(x^n\right) = nx^{n-1}$

2) a) $\frac{dy}{dx} = 2x$ **b)** $\frac{dy}{dx} = 4x^3 + \frac{1}{2\sqrt{x}}$

c) $\frac{dy}{dx} = -\frac{14}{x^3} + \frac{3}{2\sqrt{x^3}} + 36x^2$

3) *They're the same.*

4) a) $\frac{dy}{dx} = 4x = 8$

b) $\frac{dy}{dx} = 8x - 1 = 15$

c) $\frac{dy}{dx} = 2x^2 - 14x = -20$

5) *Differentiate* $v = 17t^2 - 10t$ *to give:* $\frac{dv}{dt} = 34t - 10$
so, when t = 4, $\frac{dv}{dt} = 126$ *ml/s.*

6) *The tangent and normal must go through (16, 6).*
Differentiate to find $\frac{dy}{dx} = \frac{3}{2}\sqrt{x} - 3$*, so gradient at (16, 6) is 3.*
Therefore tangent can be written $y_T = 3x + c_T$*; putting x=16 and y=6 gives* $6 = 3 \times 16 + c_T$*,*
so $c_T = -42$*, and the equation of the tangent is* $y_T = 3x - 42$*.*
The gradient of the normal must be $-\frac{1}{3}$ *so the equation of the normal is* $y_N = -\frac{1}{3}x + c_N$*; substituting in the coordinates of the point (16, 6) give* $6 = -\frac{16}{3} + c_N \Rightarrow c_N = \frac{34}{3}$*; so the normal is*
$y_N = -\frac{1}{3}x + \frac{34}{3} = \frac{1}{3}(34 - x)$.

7) *For both curves, when x=4, y=2, so they meet at (4,2).*
Differentiating the first curve gives $\frac{dy}{dx} = x^2 - 4x - 4$*,*
which at x=4 is equal to –4.
Differentiating the other curve gives $\frac{dy}{dx} = \frac{1}{2\sqrt{x}}$*, and so the gradient at (4, 2) is ¼. If you multiply these two gradients together you get –1, so the two curves are perpendicular at x = 4.*

Section Seven — Integration

1) *i) Increase the power of x by 1, ii) divide by the new power, iii) add a constant.*

2) *An integral without limits to integrate between. Because there's more than one right answer.*

3) *Differentiate your answer, and if you get back the function you integrated in the first place, your answer's right.*

4) a) $2x^5 + C$ **b)** $\frac{3x^2}{2} + \frac{5x^3}{3} + C$ **c)** $\frac{3}{4}x^4 + \frac{2}{3}x^3 + C$

5) *Integrating gives* $y = 3x^2 - 7x + C$*; then substitute x=1 and y=0 to find that* $C = 4$*. So the equation of the curve is* $y = 3x^2 - 7x + 4$*.*

6) *Integrate to get* $y = \frac{3x^4}{4} + 2x + C$*. Putting x=1 and y=0 gives* $C = -\frac{11}{4}$*, and so the required curve is* $y = \frac{3x^4}{4} + 2x - \frac{11}{4}$*. If the curve has to go through (1, 2) instead of (1, 0), substitute the values x=1 and y=2 to find a different value for C, call this value* C_1*. Making these substitutions gives* $C_1 = -\frac{3}{4}$*, and the equation of the new curve is* $y = \frac{3x^4}{4} + 2x - \frac{3}{4}$*.*

Index

A
acceleration 36
algebraic fractions 6
arithmetic progressions 32, 33
$ax^2 + bx + c$ 10

B
$b^2 - 4ac$ 15
bed of nettles 13
brackets 4, 7

C
cake 35
cancelling 7
cheerful 14
coefficients 22
common denominator 6
common difference 32
common factors 5
completing the square 12, 13, 16
constant of integration 39, 40
constants 1
cubic graphs 28
cubics 18
cup of tea 19
curve sketching 28

D
denominator 3
derivative 35, 40
differentiation 35, 36
discriminant 15
division 1
draw a picture 21
Dutchman 26

E
elimination 22
equation of a line 26
equations 1
exact answers 3
excitement 27

F
factorise 7
factorising
  cubics 18
  quadratics 10, 11
fiddly bit 11
fractions 6
functions 1

G
geometric interpretation 24
geometry club 26
godsend 14
gradients 26, 27, 35 - 37
graph transformations 29
greater than 20
greater than or equal to 20

H
hope 7

I
identities 1
indefinite integrals 39
indices 2
inequalities 20, 21
  linear 20
  quadratic 21
integration 39, 40
integration formula 39

J
James J Sylvester 11

L
laws of indices 2
less than 20
less than or equal to 20
limits of integration 39, 40
linear equation 23
lowest common multiple 22, 26

M
magic formula 14
milkshakes 25
multiplication 1
multiplying out brackets 4

N
n-shaped graphs 9, 28
natural numbers 33
normal 37
normals 27

P
parallel lines 27
perpendicular lines 27
Plato 11
points of intersection 24
polynomials 1
powers 2, 35
proof 16

Q
quadratic formula 14, 16
  proof of 16
quadratic graphs 9
quadratic inequalities 21

R
rates of change 36
rationalising the denominator 3
recurrence relations 31, 32
reflections 29
revision questions 8, 19, 25, 30, 34, 38, 41
right-angles 37
roots 2
roots of quadratics 13, 15

S
Scitardauq Gnisirotcaf 10
second order derivatives 36
sequences 31, 32
series 32, 33
sigma notation 33
simplifying expressions 4, 5, 6, 7
simplifying surds 3
simultaneous equations 22, 24
  elimination 22
  geometric interpretation 24
  substitution 23
  with quadratics 23
sketching graphs 9
splendid 20
squared brackets 4
stretches 29
substitution 23
surds 3

T
taking out common factors 5
tangent 24
tangents 36, 37
terms 1
thrill-seekers 19, 40
transformations 29
translations 29
two brackets 10

U
u-shaped graphs 9, 28

V
variables 1
vertex 9

W
washing regularly 36
wowzers 27